ORGANIC FARMING

Dr. Archana Singh, M. Sc, Ph. D, F. B. S, completed her Bachelor degree from University Maharani's College, Jaipur and Post graduate and doctorate degree from Department of Botany, University of Rajasthan, Jaipur. She has brilliant carrier in academics. She has 17 years research experience in Plant Pathology, Seed Pathology and Physiology and Biochemistry. She joined College Education in 1996 through Rajasthan Public Service Commission and has been in the profession of teaching of U.G. and P.G. classes since 1996 in Government College, Karauli, Kotputli, Bharatpur, Rajasthan. Presently she is working as the Lecturer (Selection grade) in the Government M. S. J. Post graduate College, Bharatpur. She is also engaged in guiding research scholars, working on salinity problem, plant pathology and seed pathology, and germinating wheat as an antioxidant for their Ph.D. degree. She has contributed number of research papers and articles in various National and International Journals and attended various Seminars and Conferences. She is a member of a number of professional societies.

Organic Farming

Dr. Archana Singh

Pointer Publishers
Jaipur 302 003 (Raj) India

Distributed by
Prem C. Bakliwal for
Aavishkar Publishers, Distributors
807, Vyas Building, Chaura Rasta
Jaipur 302 003 (Raj) India
Phone : 0141-2578159
e-mail : aavishkarbooks@hotmail.com

2022

ISBN 978-81-7132-647-1

Published by
Mrs. Shashi Jain for
Pointer Publishers
Vyas Building, S.M.S. Highway
Jaipur 302 003 (Raj) India
Phones : 0141-2578159
e-mail : pointerpub@hotmail.com

Price : ₹ 2700.00

Printed at
Sheetal Printers
Jaipur 302 003 (Raj) India

Dedicated
To
The Loving Memory of
My Parents
Late Shri O.P. Singh Rathore
and
Vidyawati Rathore

Foreword

On achieving self-sufficiency in food grains during green revolution through application of scientific know-how of the developed world, we happily patted our own backs with proud ignoring the contributions of all available inputs to the Indian farmers. It will be to our dismay to know that the benefits which may have accrued to the Indian agriculture have not been in consonance with the time and money spent in the last five decades. Since agriculture has been the backbone of India's economy over centuries, the ill effects of the so called advanced agricultural practices became evident to the farmers and the policy planners and the costs involved in the production of crops rose beyond the capacity of the farmers leaving only negligible share for him and his family. The soil structure and micro-flora, water quality, human health and environment have been jeopardized and not remained conducive and friendly to either human beings and or vegetative and other forms of life. With this state of affair coupled with stagnation in productivity and production, attention has been diverted to farming without the application of chemicals, i.e., fertilizers, insecticides, pesticides, fungicides, herbicides, weedicides etc. perceiving the magnitude of this debacle the advocacy of 'organic farming' came into being.

Indian agriculture has been traditionally organic since time immemorial. But in the context of the present scenario to talk about organic farming is a complicated affair. To offset some of the complexities, the Government has legislated norms which would enable one to get certify his produce as organically produced. Only time will tell about the success that may come. However, it is needless to state that organic agriculture is the only way to ensure a healthy future generation and safe guarding the biodiversity. In this context the editor of this book deserves great appreciation to have gathered chapters on different aspects of organic farming and it is hoped that the readers will find the information contained in different chapters of the book, useful, and it may enable them to understand the different facets of organic agriculture in the present scenario. As usual one always finds some shortcomings and mistakes in the publication, so is the case with this book also. But such a deficiency does not diminish the utility of the useful information so compiled and published. It is hoped that in future editions such deficiencies would be overcome.

V.B. SINGH
Former, Vice-chancellor,
Maharana Pratap University of Agriculture & Technology,
Udaipur (Rajasthan)

Preface

The emerging global prospect for Organic Farming is the production technology under agriculture, animal husbandry, poultry farming and aquaculture. It is the realization of long lasting damages caused by chemical fertilizers and pesticides in agriculture as well as for livestock, both in developed and developing countries. They penetrate in soil, water resources and disturb the ecological balances. So we can easily access and evaluate the situation and usage of chemical fertilizers and pesticides. And we can get rid of by promoting the organic farming by using various healthy and sustainable farming practices. Organic farming is the best social method for farmers, their families, livestock, farmlands and communities.

Consciousnesses about health and conservation of environment have brought a major shift towards Organic farming. Globally, people looking forward to organic food that is considered safe and hazard free. Thus, the use of synthetic fertilizers and pesticides should be avoided.

Organic farming in the form of agriculture includes crop rotation, application of green manure, compost, bio fertilizers, biological pest and diseases control to maintain soil productivity and restrict the use of synthetic fertilizers and pesticides. Infact the organic farming rely on the management of the soil. It enhances the chemical, biological and physical properties of the soil and the increasing amount of nutrition in soil plays a key role in suppressing weeds, pests and diseases in eco-friendly manner. The several compost production technologies like Vermin-composting, Phospho-composting, nitrogen rich Phospho-composting etc., improve the soil quality by enriching the nutrients.

The present book **"Organic Farming"** incorporates critical review and research articles on management of organic farming. This book contains 22 articles covering holistic information on organic farming. Important articles are on **Ancient Plant Protection Practices: Relevance As On Now, Organic Agriculture: Global and Indian Scenario, Organic Farming in India: Perceptive and Prospects, Standards and Certification and Inspection in Organic Agriculture, Organic Farming: Status and Scope in Special Reference to Indian Agriculture,** the valuable information on **Exploitation of Traditional Knowledge in the Management of**

Crop Diseases under Organic Agriculture in India: An Overview, Plant Growth Promoting Rhizobacteria for Biocontrol of Plant Diseases, Pest Management in Organic Farming, Kunapjala (Organic Manure and Pesticide) for Crop Health, detailed information is given on **Bio fertilizers: Its Production and Application in Organic Farming.** The other important articles are **Plant Growth Promoting Rhizobacteria (PGPR) Associated with Plant Nutrition, Mycorrhizal Inoculation of Potential Medicinal Plants: Field Trials and Effect on the Bioactive Phytoconstituent, Biogas Manure (BgM) a Rich Source of Plant Nutrients and Substitute to Inorganic Fertilizers, Impact of Organic Cultivation on Soil Health for Long Term Sustainability, Sustainable Crop Production through Organic Farming, its Waste and Management, Organic Culture and GM Technology, Certified Organic Farming: Evolution and Regulations,** and **Socio Economic Impact of Organic Farming: Concept, Measurement and Issues.** One article is more relevant in the present scenario which is related to human health, as pesticides and chemicals polluted the food and drinking water which cause many human diseases out of which one important disease like **Diabetes: The Chronic Health Hazard of Non Organic Farming adds value to the book.** Sufficient information is given in **Organic Live Stock Production** and **Food Production Method and Animal Health.** The text covers available information on principles and practices of organic farming related to agriculture and livestock.

I think the book will be useful and widely accepted by students, teachers, researchers, Botanists and Agriculturists. It would be of great value to all those who are interested and concerned with the organic farming.

I am grateful to all learned contributors for their cooperation in compiling useful information and update section of their articles on various aspects of organic farming by giving their valuable time to prepare such voluminous work. There may be some errors in the text after my best efforts. I welcome the comments and suggestions of readers and reviewers.

I wish to offer my heartiest appreciation to my husband Dr. Rajvir Singh whose patience, keen interest, understanding and constant support throughout the preparation of this volume. Sincere appreciation is expressed to my little son and daughter master Pratyush and Shambhavi and my niece Vijaylakshmi for their valuable cooperation and environment at home to complete this task.

I also record my sincere gratitude to Mr. Vipin Jain of Pointer Publishers, Jaipur for publishing this book with patience and interest.

Bharatpur **—Archana Singh**

Contents

Chapter 1

Ancient Plant Protection Practices : Relevance as on Now

S.L. CHOUDHARY AND GANESH RAJAMANI

AAHF Rajasthan Chapter, 105, Vidya Nagar, Hiran Magri, Sect.4, Udaipur, Rajasthan

References to plant protection are found in Vedas (Rigveda c.3700 BC Atharvaveda c.2000 BC), Kautilya's Artha-sastra (c.300 BC), Buddhist literature (c.200 BC), Krishi Parashar(c.100 BC), Sangam literature of Tamils (200 BC-100 AD), Agnipuran (c.400 AD), Brhat Samhita of Varahamira (c.600 AD), Kashyapiyakrisukti (c. 800-900 AD), Surapala's Vrikshayurveda (c.1000AD, Someshwara Deva's Manasollasa (c, 1100A D), Lokopakara) by Chavundaraya (c.1108 AD), Sarangadhara's Upavanavinoda (c.1300 AD),Viswavallabh of Chakrapani Mishra (c.1577 AD), and some documents of the medieval and pre-modern period. But Surapala c.1000 AD) has given plant protection in a very systemic manner right from seed treatment to the storage of grains. Therefore, this period may be considered as the starting point of systematic plant protection in Indian agricultural history.

AILMENTS DESCRIBED BY SURAPALA

Diseases of all kinds of trees are stated to be of two types: internal and external. It is unfortunate that all textbooks on plant pathology give credit to the French botanist, Tournefort (1075 AD) for classifying diseases as internal and external. This was more than 700 years after Surapala had already done such classification. For the 'internal disorder of plants he borrowed the *tridosha* principle of Ayurveda, classified "internal causes as the imbalance of humors, *vata*, *kapha* and *pitta*; and external ones are caused by insects, cold weather, etc. (for details please read the chapter Plant Protection in "A Textbook on Ancient History of Indian Agriculture" by R C Saxena, S L Choudhary and Y L Nene, 2009).

TREATMENTS OF AILMENTS SUGGESTED BY SURAPALA

Diseases caused by *vata* can be cured by flesh, marrow, and ghee; sprinkling of *kunapa* water and liberal fumigation of the mixture of fat of hog, oil of Gangetic porpoise, ghee, hemp, hair of horses and cow's horn-all boiled and set to a decoction. Likewise diseases caused by *kafa* humor can be cured with bitter, strong and astringent decoction made out of *Panchamula* (roots of five plant species-*sriphala, sarvatobhadra, patala, ganikarika* and *synoka*) with fragrant water or the paste of white mustard should be deposited at the root and trees should be watered with a mixture of sesame and ashes. In case trees are affected by the *kafa* disease, soil around the roots of the tree should be removed and fresh dry soil should be replaced for curing them. To get cured from *pitta* type of diseases, treat trees with cool and sweet substances-when watered by decoction of milk, honey, *yastimadhu* (*Glycyrrhiza glabra* Linn.) and *madhuka* (*Madhuka indica* J. F. Gmel). Further, when watered with the decoction of fruits of *triphala* (the three kinds of myrobalan-*Terminalia chebula* Retz, T. *belliraca* Roxb., and *Phyllanthus emblica* Linn.), ghee, and honey the trees are freed of all diseases arising from the state of ailments.

To remove insects both from the roots and branches of the trees, water the trees with cold water for seven days. The worms population can be over come by the paste of *kunapa jal* (water) and cow dung mixed with water and also by smearing the roots with a mixture of white mustard, *vaca* (*Zingiber zerumbet* Rosc. Ex Smith.), *kusta* (*Saussurea lappa* C. B. Clarke), and *ativisa* (*Aconitum heterophyllum* Wall ex Royle). Likewise the insects on the leaves can be destroyed by sprinkling the powder of ashes and dust. A wound caused by insects heals if sprinkled with milk after being anointed with a mixture of *vidanga*, sesame, cow's urine, ghee, and mustard. Other wounds of the trees are healed by the treatment of anointing with the paste of bark of *nyagrodha* (*Ficus bengalensis* Linn.), and *udumbara* (*F. glomerata* Roxb.), cow dung, honey, and ghee. The oozing can be cured by the use of above paste and covering the part with the bark of *dhava* (*Anogeissus latifolia*), *sriparnika* (*Myrica esculenta* Buch-Ham.), *syama* (*Ichnocarpus frutescens* R. Br.), *vetasa* (*Salix capera* Linn.) and *arjuna* (*Terminalia arjuna* (Roxb.) Wight & Arn.

Similarly Surapala had suggested the different treatments like sprinkling of *kunapa* water and milk, anointing the branches of trees with *vidari* (*Pueraria tuberose* DC), sugar, *nagajivha* (red arsenic) and sesame mixed together and sprinkled with milk water to the trees suffering from frost, scorching heat; and if branches burnt , respectively. Further, if the trees dried due to bad soil, the original soil from the root should be removed and it should be replaced by healthy soil and milk-water should be sprinkle on it. If the drying is due to lack of water the trees should be watered with milk-water and properly fomented by smoke of the crab shells.

Diseases caused by wrong treatment can be conquered by sprinkling the mixture of water and milk and also by applying a paste of *vidinga* mixed with thick mud. Jaundice (yellowing) can be brought under control only in weeks by sprinkling water mixed with the powder of barley and wheat added to honey and milk.

INFECTIOUS DISEASES

Surapala suggested that before planting cuttings in the pits, the latter should be 'burned' using dry plant material, cow dung etc. This is an indication of a suspicion that Surapala must have known about existence of infectious entities. Chakrapani Mishra (1577

AD) suggested that diseased plants found in the midst of healthy plants should be removed and burned, this again pointing towards existence of infectious entities. It is unfortunate that all the current textbook on plant pathology credit Tillet, who in 1755 AD dusted wheat seed with 'bunt' spores to produce the disease called wheat bunt. We should, however know that Koch's postulates have to be followed to prove infectious nature of a disease, here again, Indians have not been given due credit by the authors of the West.

Vishavavallabha is an another treatise written by Chakrapani Mishra under the patronage of Maharana Pratap of Mewar on the science of plant life which resembles Surpala's Vrikshayurveda and deals more or less with the same subject but with some additions. For example, several new herbs have been mentioned for the control of disorders, such plant species are, *ambu* (*Pavonia odorata* Willd.), *aragavadha* (*Cassia fistula* L.), *arishta* (*Sapindus emarginatus* Vahl.), *ingundi* (*Nalanites aegyptiaca* (L.) Delile), *karanja* (*Pongamia pinnata* (L.) Pierre), *katphala* (*Myrica esculenta* Buch.-Ham, ex D. Don), *katvanga* (*Ailanthus excelsa* Roxb.), *kuberakshi* (*Caesalpinia crista* L.), *nimba* (*Azrdirachta indica* A, Juss.) bark, *rohita* (*Tecomella undulate* (Smith) Seem.), *shatapushpa* (*Anethum sowa* Kurz.), *tagara* (*Valeriana jatamansi* Jones, *vasa* (*Adhatoda vasica* Nees), etc.

Apart from Vrikshayurveda and Vishvavallabha paramount documents concerning plant protection were Someshwara Deva's Manosollasa (1131 AD), Sarangdhara's Upvanvinoda (1300 AD), Bhavprakash-nighantu (1600AD), Tuzak-i-Jahangiri (1605 AD), Dara Shikoh's Nuskha Dar-Fanni-Falahat (1650 AD), Jati Jai Chand Diary (1658-1714 AD), an anonymous Rajasthani manuscript from Mewar region of Rajasthan(1877 AD), and Watt's Dictionary of Economic Products of India (1889-1893).

Jahangir, the Mughal Emperor of India (1605-1627) described in his memoir "a disorder of marigold" which could be ascribed today to species of *Alternaria, Botrytis* or *Sclerotium*. Similarly in Jati Jaichand diary the early blight (*Curvularia penniseti*) of pearl millet and possibly Botrytis gray mold of chickpea have been described.

In a document of early 19th century from the Mewar region of Rajasthan, powdery mildew has been described infesting various plants alongwith canker or anthracnose of orange. In this document a number of plant protection practices have been given. Some interesting practices are:

1. Use of oil (probably sesame) for soil and foliar application to trees to protect from frost and termites.
2. Sprinkling of curd (9 L) with asafetida (112 g) on trees to prevent powdery mildew.
3. Use of asafetida and *vidanga* mixed with curd every 10 days to protect against orange canker.

Use of cow dung for smearing the cutting of fig before planting is mentioned in Dara Shikoh's Nuskha Dar Fanni_Falahat (1650 AD). Garlic has been mentioned specially for insect control. In addition to these he has mentioned the use of salt solution for soaking fig cuttings before planting. This is followed by cow dung application.

INDIGENOUS PLANT PROTECTION PRACTICES STILL FOLLOWED

In traditional agricultural practices, farmers evolved in an effective system of crop protection through generations of experience and intimate knowledge of their environment which are still followed by the farmers in different parts of the country. Some of such practices are:

Application of organic manure, summer ploughing, crop rotation, use of *neem* leaves and *neem* cake, seed treatment with ash, etc.

Besides these several indigenous practices which are still followed by the farmers of Arunchal Pradesh and Rajasthan are mentioned below:

1. **Buttermilk for pest control**. Farmers treat garlic seeds with buttermilk @ 10-12 L/ha before sowing for protection against different diseases and insect pests.
2. **Asafetida to control of ergot of sorghum**. About 50-60 g of asafetida in water is used for making solution to treat sorghum seeds for protection against ergot.
3. **Control of ginger rot and yellowing.** The tribal farmers sow ginger after treating with a solution of cow dung. About 10-15 kg cow dung and asafetida (40-80g/) are mixed in 8-10 L of water. This is enough to treat ginger rhizomes required for one hectare. This practice is helpful for protecting ginger plants from rotting during vegetative growth. The farmers grow papaya as an intercrop for providing shade to ginger and to prevent yellowing. The leaves of ginger do not turn yellow and high yields are obtained.
4. **Control of tomato wilt**. Farmers use turmeric powder @ 15 to 20 kg/L of water to control wilt (*Fusarium* sp.) of tomato. This solution is used to treat the roots of seedlings grown in nurseries of tomato before transplanting.
5. **Control of early stem borer in sugarcane.** To control the stem borer (*Chilo infuscatellus*) in sugarcane, farmers use *neem* oil @ 1.5-2 L/ha. This practice is repeated thrice in the whole period of sugarcane vegetative growth.
6. **Control of termites.** Farmers keep asafetida in a pack of cotton cloth at two or three points in the irrigation channel of 10-15 m irrigation for controlling termites (*Odonotermes obescus*) in affected crops.
7. **Control of storage pests.** Farmers used different simple practices for the control of storage pests viz. fuel wood ash for pests of pulses (250 g ash/250 kg of pulses); dry *neem* leaves @ 2.5 kg/100 kg of wheat for controlling storage pests, dry chillis @ 5-6 per kg of seed of mung bean and black gram.

PLANT PROTECTION IN RAJASTHAN—SEMI-ARID REGION

In Rajasthan some of the following methods of plant protection in semi-arid and tribal regions are used by farming communities.

Control of Termites

1. Attack of termites and cracking of epidermis in fruits can be avoided by using a paste on the main trunk with a mixture of mustard oil (one l)+turmeric powder (200 g). Mustard oil attracts ants which remove termites while turmeric powder helps in healing the cracks. Another practices is chopped the *aak* (*Calotropis*) leaves, filled in gunny bags and kept in irrigation channel. The exudates from these leaves have been reported to kill termites and other soil-borne pathogens.
2. Farmers amend the soil with the pre-incubated mixture of *aak* leaves (5 bags)+*neem* kernels and leaves (95 bags)+asafetida (200 g) to check termite attack

in the field and soil-borne diseases. They claimed good growth of crop due to this amendment.

Control of Thrips, Aphids, Whitefly and other Pests as well as Virus Diseases

1. To prevent thrips and aphids they apply cow dung ash on foliage of vegetables e.g., chilli, onion, garlic, and cucurbits
2. In Jodhpur district some farmers use cow urine based bio-pesticide to check whitefly, jassids, and other sucking pests of chilli and cumin. This is prepared by mixing cow urine (10 L)+*neem* kernels and leaves (2 kg)+garlic (100 g).The ingredients are crushed and kept in copper container for 15 days. The solution is then heated till 5 L of solution is left.

After cooling, the liquid is filtered and diluted with 500 L of water before spraying. The farmers even claimed its efficacy in controlling viral diseases of chilli also.

Control of nematodes. Mixture of cow dung+*aak* (*Calotrips* spp.)+*kheip* or *khip* (*Crotalaria burhia*), local xerophic plant foliage is allowed to rot in a pit for about two months. This mixture (manure) is then applied in chilli and tomato fields for control of root-knot nematodes, termites as well as for good growth.

Tribal Region

1. In paddy, spraying a solution of 4 L of cow urine and 10 g asafetida in 10 L of water repels the sucking pests (aphids and jassids).
2. To control paddy blast and bacterial blight, spraying a solution of cow dung prepared by mixing 3 kg cow dung in 3 L of water is commonly used by the farmers.
3. In case of insects holes made by shoot borer and bark eaters in mango trees, jaggere is placed in the holes to attract other predators (ants), which will feed on the insects present in the holes. Similarly the practice of pouring kerosene oil in holes made by insects and blocking holes with cow dung is also popular.
4. A peculiar method of controlling pathogen attack in chilli is practiced by tribals wherein the twigs of *aak* (*Calotropis* spp.) are placed in between rows. Similarly some farmers placed fresh cow dung near the collar region of chilli plant to prevent it from fungal diseases viz., damping off and die back.
5. In case of soil-borne diseases viz., root rot, collar rot, etc. and termites, the cakes of castor, *karanj* or *neem* are used as control measures.
6. During sprouting of sets in sugarcane crop, putting stem of *aak* (*Calotropis* spp.) in the irrigation channel during intercultural operations has been found effective against termites.

SCIENTIFIC BASIS OF AGE-OLD PLANT PROTECTION PRACTICES

It has been now realized that the techniques adopted for commercial agriculture are unsustainable on long term basis. Therefore, agricultural scientists are diverting their attention to the traditional or indigenous technology and exploring possibilities of using them wherever possible. Our old traditional technologies were scientific and almost eco-friendly as all the plant protection practices were based on organic materials both of plant and animal origin which includes honey, ghee, milk and milk products, cow dung and urine, and extracts from number of plant species like *Brassica* ssp., *Madhuka indica*, *Ficus* spp., *Piper nigrum*, *Azadirachta indica*, *Vitex nigundo*, *Embelia ribes* etc. The biochemical analysis of these materials clearly indicated now that all these materials have antimicrobial activities.

Milk and ghee have been used for centuries. Even buttermilk was found useful. About 40% of total amino acids in milk are glutamate, leucine, and proline. A recent report claimed that milk sprays induced systemically acquired resistance in chilli against leaf curl, a viral disease. Milk (10% acqueous suspension) has also been effectively used for controlling powdery mildews.

The use of cow dung by our farmers for different purposes like seed dressing, plastering cut ends of vegetatively propagating units such as sugarcane sets, dressing wounds, sprinkling dilute suspension of plants and applying to soil has been indicated since the time of Kautilya (c.300 BC). The cow dung from the cattle shed is a mixture of dung and urine in a ratio of 3:1. Cow dung consists crude fibre, crude protein and materials that can be obtained in nitrogen-free extracts. There are more than 60 species of bacteria and 100 species of protozoa encountered in the rumen of cow. Thus when seed is treated in various ways with cow dung, it gets coated with cow dung residue. The residue contains several organic elements, enzymes, macro and micro-nutrients, epithelial cells, bile salt and pigment and large number of bacteria. The dung residue has emulsifying properties and readily absorbs moisture from the surrounding soil to the advantage of seeds. The presence of bacteria may antagonize potential pathogens ready to attack seed.

Neem cake application to the field reduces population of soil-borne fungi and nematodes and also reclaims alkaline soil due to presence of calcium and magnesium. The ancient practice of spreading of *neem* leaves over groundnut in storage has a scientific basis. It has now proved that *neem* leaves inhibit the growth of *Aspergillus flavus* and thereby prevent aflatoxin production.

The utility of *neem* tree has been recognized long back in Indian agricultural history. Every part of this tree is used for a number of purposes. Now this tree is found to be an effective air filter and protects environment.

❑❑❑

Chapter 2

Organic Agriculture : Global and Indian Scenario

B.S. Mahapatra*, T. Ramasubramanian*, H. Chowdhury*, S.P. Singh* and D.K. Singh**

**Central Research Institute for Jute and Allied Fibres (ICAR), Barrackpore, Kolkata - 700 120*
***Department of Agronomy, G.B. Pant University of Agriculture and Technology, Pantnagar – 263 145*

INTRODUCTION

The food grain production of India has touched the land mark figure of 233.88 million tonnes in 2008-09 (Economic Survey, 2009-10) from mere 95 million tonnes in 1967-1968 with virtually no increase in net cultivated area of 140±1 m ha. Green revolution of India has undoubtedly changed the scenario of food grain production from 'ship-to-mouth' status to self sufficient and surplus status. This marvellous achievement in agricultural production was mainly due to the increased use of inputs like fertilizers, pesticides and farm machinery. However, the achievement was at the expense of ecology and environment, which is detrimental to the well-being of the people. Though the modern farming system adopted from the west had helped India to attain the self sufficiency in food grain production, it has increased the un-sustainability. The land is losing its fertility and is demanding larger quantities of fertilizers. Pests are becoming resistant and warrant frequent sprays, which resulted in residues much above the safety levels. The necessity of having an alternative eco-friendly agricultural method, which can sustain and / or increase the crop productivity, has been realized. Organic farming has been recognized as the best known alternative to the conventional or chemical agriculture. The practice of organic farming also originated in the west, which suffered from the ill-effects of chemical agriculture. Organic agriculture aims at the human welfare without any harm to the environment. The relatively high success of organic farming in some countries are due to high awareness about health problems caused by the consumption of contaminated food products, the ill-effects of environmental degradation, appropriate policy measures by the governments and organizations like the European Union and International Federation of Organic Agriculture Movements (IFOAM). Further stringent standards for non-organic food in European and US markets have led to the rejection of many Indian food consignments in the recent past and this brought to the attention of ill-effects of modern agriculture, and paved the way for organic farming. Organic farming has entirely different perspective between developed and developing countries. Developed countries are endowed with surplus resources, but limited population with high purchasing power. Hence, a consumer in a developed country expects enough food of highest possible quality that has been produced without any serious ecological impact and harmful residues. Thus organic farming may be the best answer to fulfil such expectations in rich countries (Marwaha and Jat, 2004). However, on the other hand, for a resource poor, highly populous country like India, with ever growing demand for food, the primary concern may be enough availability of food at affordable prices with reasonable quality. In this backdrop, considerable efforts have been taken to promote the organic farming in India without compromising the food grain production to feed the ever growing population (Mahapatra *et al.*, 2009).

ORGANIC AGRICULTURE: GLOBAL SCENARIO

Status of Organic Agriculture Across the Continents

A little more than 32 million ha, which accounts for 0.78% of total agricultural land has been brought under the organic agriculture across the globe by 2007. The continent Oceania, which includes Australia, Fiji, New Zealand, Niue, Papua New Guinea, Samoa, Solomon Islands and Vanuatu ranks first in terms of percentage of total arable land under organic agriculture. In Oceania the area under organic agriculture has been worked out to be little more than 12 million ha, which accounts for 2.61% of the total arable land of the continent. Australia alone accounts for 12.02 million ha out of the total 12.11 million ha of the Oceania under organic agriculture. This was followed by the European continent with 1.87% (7.62 million ha) of the total agricultural land under organic agriculture. In Asia, organic agriculture is practiced in 2.9 million ha accounting 0.21% of the total arable land of the continent. The share of organic land out of the total arable land has been reported to be 0.1%, 0.56% and 1.02% for Africa, North America and Latin America, respectively (Table 1) (Source: FiBL and IFOAM, 2009). The area under organic agriculture in the world has been increased gradually from 14.78 million ha in 2000 to 32.09 million ha in 2007. The area has been increasing gradually over the same period in all the continents (Fig. 1). The area under organic agriculture in Asia has been increased by 46.5 times from mere 0.06 million ha in 2000 to 2.9 million ha in 2007. The area under organic agriculture has been increased by 16.64, 2.27, 2.07, 1.76 and 1.63 times in Africa, Oceania, North America, Europe and Latin America, respectively (Table 2) (Source: FiBL and IFOAM, 2009).

Table 1: Area under organic agriculture in different continents

Continent	Area (ha) under organic agriculture	%organic of area under organic agriculture
Africa	875,370	0.10%
Asia	2,900,068	0.21%
Europe	7,627,915	1.87%
Latin America	6,380,996	1.02%
North America	2,197,042	0.56%
Oceania	12,110,758	2.61%
Total	32,092,149	0.78%

(*Source:* FiBL and IFOAM, 2009)

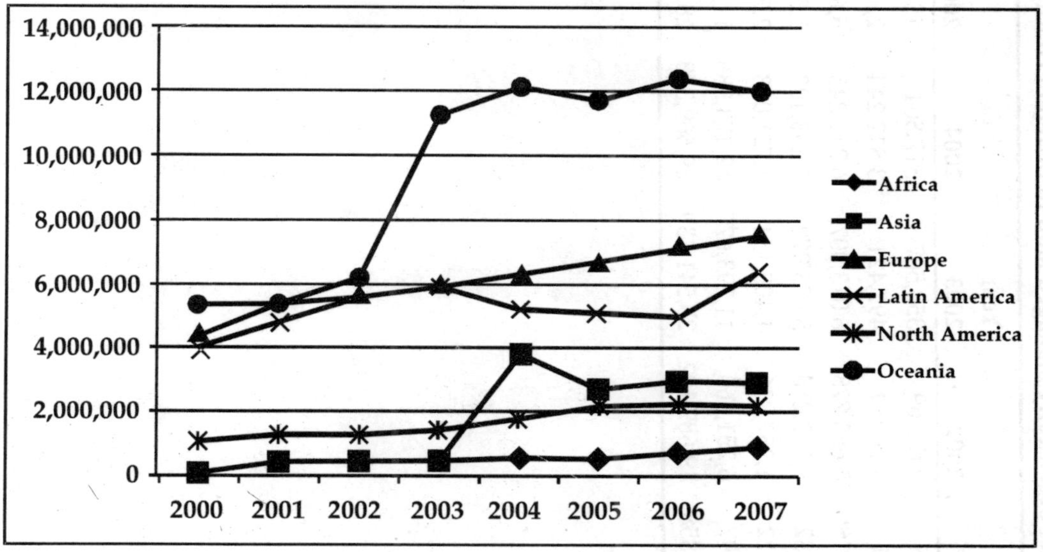

Fig. 1: Increase in area (ha) under organic agriculture in continents over the years (Source data: FiBL and IFOAM, 2009)

Status of Organic Agriculture Across the Countries

As of 2007, Australia ranked first in terms of area under organic agriculture (12.02 million ha). This was followed by Argentina (2.78 million ha), Brazil (1.77 million ha), USA (1.64 million ha), China (1.55 million ha) and Italy (1.15 million ha). India with 1.03 million ha under organically managed land, ranked 7th among the 138 countries practicing the organic agriculture (Table 3). Twenty countries have more than 5% the total agricultural land under organic agriculture. The percentage share of organically managed land to the total agricultural land was the highest in Liechtenstein (29.7%), followed by Austria (13.4%) and Switzerland (11.0%) with more than 10% of the total arable land under organic agriculture (Table 4). The percentage share of organically managed land to the total agricultural land in India was 0.6% in 2007.

Table 2 : Area under organically managed land in different continents over the years

Continent	Year							
	2000	2001	2002	2003	2004	2005	2006	2007
Africa	52,601	233,558	317,169	358,507	517,804	486,674	684,978	875,370
Asia	62,330	421,381	424,289	497,421	3,812,244	2,682,182	3,000,607	2,900,068
Europe	4,345,887	5,296,918	5,683,778	6,013,007	6,211,245	6,668,395	7,207,209	7,627,915
Latin America	3,917,676	4,767,682	5,752,270	5,973,972	5,233,402	5,072,158	4,966,295	6,380,996
North America	1,058,951	1,278,122	1,257,936	1,405,154	1,721,063	2,219,643	2,245,173	2,197,042
Oceania	5,344,197	5,344,197	6,251,660	11,293,677	12,173,586	11,811,868	12,431,820	12,110,758
Total	14,781,642	17,341,858	19,687,102	25,541,739	29,669,343	28,940,920	30,536,082	32,092,149

(*Source:* FiBL and IFOAM, 2009)

Table 3 : Countries with more than one million hectares under organic agriculture (fully converted + area under conversion)

Rank	Country	Area (ha)
1	Australia	12,023,135.1
2	Argentina	2,777,959.0
3	Brazil	1,765,793.0
4	USA	1,640,769.2
5	China	1,553,000.0
6	Italy	1,150,253.0
7	India	1,030,311.0

(*Source*: FiBL and IFOAM, 2009)

Table 4 : Countries having more than 5% the total agricultural land under organic agriculture

Rank	Country	Share of organically managed land to the total agricultural land (%)
1	Liechtenstein	29.7
2	Austria	13.4
3	Switzerland	11.0
4	Sweden	9.9
5	Italy	9.1
6	Estonia	8.8
7	Latvia	8.5
8	Samoa	7.8
9	Czech Republic	7.4
10	Timor Leste	7.0
11	Finland	6.5
12	Dominican Rep.	6.3
13	Uruguay	6.2
14	Vanuatu	6.1
15	Slovenia	6.0
16	Portugal	5.9
17	Slovak Republic	5.6
18	Denmark	5.4
19	Germany	5.1
20	Sao Tome and Principe	5.0

(*Source*: FiBL and IFOAM, 2009)

ORGANIC AGRICULTURE : INDIAN SCENARIO

India is bestowed with lot of potential to produce organic products due to its various agro-climatic regions. The inherited tradition of organic farming is an added advantage. The cultivated land under certification in India has been worked out to be 2.8 million ha during 2007-08, of which one million ha was under cultivation and the rest was under forest area (wild collection) (Source: Agricultural and Processed Food Products Export Development Authority - www.apeda.com). Madhya Pradesh has 1.47 lakh ha of area under organic certification followed by Maharashtra (1.13 lakh ha), Orissa (74,585 ha), Jammu and Kashmir (32,541 ha), Rajasthan (24,731 ha) Mizoram (16,802 ha) and Kerala (14,744 ha) (Table 5).

Table 5 : Area under organic certification in different states of India (2006-07)

Rank	State	Area in hectares		
		Certified area	Under conversion	Total
1	Madhya Pradesh	87536.0	59875.8	147411.8
2	Maharashtra	41390.5	72238.4	113628.9
3	Orissa	66625.4	7959.7	74585.1
4	Jammu and Kashmir	32541.8	0.0	32541.8
5	Rajasthan	15034.3	9697.5	24731.8
6	Mizoram	0.0	16802.5	16802.5
7	Kerala	11631.9	3112.7	14744.7
8	Karnataka	8735.1	2976.8	11711.8
9	Andhra Pradesh	5561.2	4925.9	10487.1
10	West Bengal	7332.8	3147.2	10479.9
11	Himachal Pradesh	69.0	9507.7	9576.7
12	Uttarakhand	3178.6	5250.9	8429.5
13	Gujarat	7102.3	658.5	7760.8
14	Uttar Pradesh	5589.6	1700.6	7290.1
15	Goa	4100.5	2849.8	6950.3
16	Manipur	913.7	5105.9	6019.6
17	Delhi	3632.6	1830.4	5463.0
18	Tamil Nadu	3414.1	1652.4	5066.5
19	Haryana	3382.5	15.8	3398.3
20	Assam	2526.6	540.2	3066.9
21	Jharkhand	10.5	2253.4	2263.9
22	Sikkim	274.8	1531.9	1806.7
23	Other	510.5	966.3	1476.8
24	Punjab	347.6	698.4	1046.0
25	Nagaland	0.0	878.9	878.9

contd. ...

... contd.

Rank	State	Area in hectares		
		Certified area	Under conversion	Total
26	Arunachal Pradesh	65.9	632.8	698.6
27	Chhattisgarh	279.2	28.7	307.9
28	Meghalaya	0.0	304.4	304.4
29	Bihar	0.0	0.0	0.0
30	Tripura	0.0	0.0	0.0
	Total	311786.9	217143.4	528930.3

GLOBAL ORGANIC MOVEMENT

At global level, the roots of organic farming can be traced back to the first quarter of early 20th century in Europe (Stockdale *et al.*, 2001). In 1924, the Austrian philosopher Dr. Rudolf Steiner conceptualized and advocated organic agriculture and in 1927 a trademark 'Demeter' was introduced for organically produced food. Organic movement began in the 1940s as a reaction to agriculture's growing reliance on synthetic fertilizers and chemical pesticides, leading to the decade being referred to as the "pesticide era". During this period, path breaking literature on the subject published by J.I. Rodale in the United States, Lady Balfour in England and Sir Albert Howard in India contributed to the cause of organic farming. Sir Albert Howard is widely considered as the father of organic farming. The formation of the International Federation of Organic Agriculture Movements (IFOAM) in 1972 gave framework for the discussion and codification of internationally recognized principles of organic farming. The main principles of organic farming are as follows:

- To produce food of high quality in sufficient quantity.
- To interact in a constructive and life-enhancing way with natural systems and cycles.
- To consider the wider social and ecological impact of the organic production and processing systems.
- To encourage and enhance biological cycles within the farming system, involving micro-organisms, soil flora and fauna, plants and animals.
- To maintain and increase the long-term fertility of soils.
- To maintain the genetic diversity of the production system and its surroundings, including the protection of wildlife habitats.
- To promote the healthy use and proper care of water, water resources and all life therein.
- To use, as far as possible, renewable resources in locally organized production systems.
- To give all livestock conditions of life with due consideration for the basic aspects of their innate behaviour.
- To progress towards an entire production, processing, and distribution chain, which is both socially justifiable and ecologically responsible.
- To minimize all forms of pollution.

ORGANIC MOVEMENT IN INDIA

Organic farming has been practiced in India since thousands of years. The great Indian civilization thrived on organic farming and was one of the most prosperous countries in the world. In traditional India, the entire agriculture was practiced using organic techniques, where the fertilizers, pesticides, etc., were obtained from plant and animal products. At that time, organic farming was the backbone of the Indian economy. The country had witnessed a tragic Bengal famine in 1942-43 that claimed millions of lives due to starvation. India's annual food grain production in 1950-51 was only about 50 million tonnes. In this backdrop, modern agriculture played a vital role in alleviating the hunger through green revolution. India's achievement in agricultural production after the green revolution has been exemplary and mainly due to the increased use of inputs like fertilizers, pesticides and farm machinery. The food grain production in India itself more than doubled during the post green revolution period with virtually no increase in net cultivated area. This practice of intensive modern agriculture for the past five decades has led to several environmental and health issues. The land is losing its fertility and is demanding larger quantities of fertilizers. Pests are becoming resistant requiring the farmers to use stronger and costlier pesticides. The over-use of pesticides resulted in residues much above the safety levels and this brought to the attention of ill-effects of modern agriculture, and paved the way for organic farming. According to the Agricultural and Processed Food Products Export Development Authority (APEDA), around 2.8 million hectares of land was under organic farming in India in 2007. Further, there are over 15,000 certified organic farms in India. India is, therefore, one of the most important suppliers of organic food to the developed nations. No doubt, the organic movement has again started in India.

ISSUES AND OPPORTUNITIES FOR ORGANIC FARMING IN INDIA

The agriculture and allied sectors in India provide employment to 65% of the workers and accounts for 30% of the national income. Population growth and the increase in income will lead to a rise in demand for food grains and also for the agricultural raw materials for industry in the future. The area under cultivation, obviously, cannot be increased and the present 140 million hectares will have to meet the future increases in demands. There is a strong reason for even a decline in the cultivated area because of the urbanization and industrialization, which in turn will exert much pressure on the existing cropped area.

According to Marwaha and Jat (2004), the organic farming is probably seldom practised and there is dearth of data on comparative yield and economics of crops with pure organic farming versus intensive farming. Whatever little organic farming that is being practised in India seems to be taken up mainly as a part of contract farming. In India, the area under organic farming is estimated to be only 0.6% of total cultivated area. Even in Europe and other developed countries of the world, despite serious efforts and incentives, the area under organic farming is generally less than 5%. Organic farming may be far from reality for three reasons. Firstly, the limited availability of organic sources; secondly, for getting optimum yields, crops require nutrients in sufficient quantities and in readily available form. This could be best possible through fertilizers, and hardly through organic sources. Thirdly, for plant protection, the organic sources and biological methods that have been tried so far have not met with considerable success in the country. Marwaha and Jat

(2004) were of the opinion that, organic farming may be promoted in selected rainfed areas and for export oriented crops in India.

In India, only about 38% of the cultivated land is irrigated and in rest of the rainfed areas (62%) there is little or no use of fertilizers and other agrochemicals on account of associated risk in farming and poor resources. Thus, almost 62% of such areas of the country may be regarded as a kin to organic farming. Therefore, there could be ample scope for pure organic farming in those areas. Similarly, some areas under tea, coffee, spices and cashew may be easily brought under organic farming with a thrust on export of organic produce. Thus, rather than promoting organic farming *en masse* and universally, it would be desirable to carefully delineate areas for organic agriculture, which provides better profitability and boost to the export (Marwaha and Jat, 2004).

Several states in India had already initiated steps to encourage the organic farming among the farming communities. Among the state governments, Madhya Pradesh took early lead in organic farming followed by Uttaranchal (declared as organic state) and other states. The 'Organic Farming Council of Punjab' established in 2006 is taking lead in promoting organic farming in Punjab. Green manuring with *Sesbania/Crotolaria* – Pusa Basmati – Chickpea cropping system under organic mode has been identified as a complete rotation to be followed in *Tarai* region of Uttarakhand. Wheat, lentil and mustard were some of the important crops successfully grown under organic mode with yield realization of 25-27 q, 12-14 q and 10 q, respectively at Pantnagar. There are certain researchable issues need to be addressed for popularization and subsequent successful implementation of organic farming in India. The issues are:

- Suitable land races / local cultivars of high value crops need to be identified for organic cultivation.
- Improved methods need to be standardized for enriching the nutrient status of the composts.
- Different cropping sequences need to be evaluated under various agro-ecological situations.
- The economics of compost preparation, cropping systems and seed production need to be worked out under various farming systems.
- Soil health, weed, pests and disease incidence need to be monitored regularly.
- The possibilities of incorporating location-specific livestock components in organic farming need to be studied.
- The indigenous technical knowledge (ITKs) need to be documented for further exploitation in organic farming.

CURRENT STATUS OF REGULATORY MECHANISMS

At Global Level

Organic agriculture is based on the commitment of farmers and processors to work according to a set of standards and regulations, which define the organic production system. Further, organic agriculture is based on transparency to make the production system comprehensive and reliable. Finally, organic agriculture needs the consumer's confidence for market development. Certification and accreditation systems primarily serve as tools to enhance trade, market development and foster confidence among the consumers (Herrmann and Rundgren, 2007). The first international standards for organic products were published by IFOAM in 1980. The first legislative initiatives were developed by some European countries (e.g. Austria, France) in the 1980s. In 1991, the EU passed the organic regulation 2092/91 and set standards with major implications for international trade, and included not only production standards, but also standards for labelling and inspection. Various countries in Europe, Latin America and Asia introduced legislation in the 1990s. In the new millennium, most major economies have implemented legislation on organic production. Currently more than 60 countries have a regulation. In India the regulations for organic agriculture has been implemented only for export commodities (Huber *et al.*, 2007).

As of 2006, around 395 organizations worldwide offer organic certification services. Most certification bodies are in Europe (160), followed by Asia (93) and North America (80). Many of the certification organizations also operate outside of their home country; 40% of the certification bodies are approved by the European Union, 32% have ISO 65 accreditation, and 28% are accredited under the US National Organic Program (Herrmann and Rundgren, 2007).

Lack of acceptance and recognition among the different certification and accreditation systems can contradict the objective of enhancing trade, market development and fostering confidence. An important initiative for international harmonization is the IFOAM Accreditation Program, which assesses certification bodies against the IFOAM norms. The International Organic Accreditation Service (IOAS) is a non-profit independent organization registered in Delaware, USA that offers international oversight of organic certification, through a voluntary accreditation process for certification bodies active in the field of organic agriculture. The IOAS implements the IFOAM Accreditation Program. Under this program, applicant certification bodies are assessed against the IFOAM Norms (Herrmann and Rundgren, 2007).

At National Level

National Programme for Organic Production (NPOP)—There is tremendous scope for cultivation and export of wide range of organic produce in India. Looking into its importance, Ministry of Commerce, Government of India launched National Programme for Organic Production (NPOP) in 2000. Declaring standards for organic production, policy on accreditation system, inspection and certification process and institution of logo were the major objectives of this programme at the time of initiation. A national organic logo "India Organic" has also been developed that has been in use for certified organic products. The NPOP standards for production and accreditation system have been recognized by European Commission and Switzerland as equivalent to their country standards. Similarly, USDA has also recognized NPOP conformity assessment procedures of accreditation as equivalent to that of USA. With these recognitions, Indian organic products duly certified by the accredited certification bodies of India (Table 6) are well-accepted by the importing countries.

Accreditation agencies—As of 2004, Agricultural and Processed Food Product Export Development Authority (APEDA), Coffee Board, Spices Board, Tea Board, Coconut Development Board, and Cocoa and Cashewnut Board were authorised by the Ministry of Commerce, Government of India as accreditation agencies for organic agricuture (Bhattacharyya and Gehlot, 2004).

Certifying agencies—At present there are 18 accredited certification bodies in India (www.apeda.com as on 13.05.2010) (Table 6). In 2001, a group of organizations and corporate bodies took the initiative to set up the Indian Organic Certification Agency, INDOCERT. It has become an important element of the organic movement in India and mobilizes new forces and partnerships. In 2003, together with other partners, INDOCERT created the International Competence Center for Organic Agriculture (ICCOA), a service provider for networking, capacity building and market development in the organic sector in India. Among other activities, ICCOA implements the Indian Organic Market Development Project (2005-2007), the activities of which were focussed on awareness raising, market intelligence, developing organic market initiatives, and the India Organic Trade Fair (Kilcher, 2007).

Table 6 : List of accredited certification bodies under NPOP

Name of the certification agency	Validity of current accreditation	Scope of accreditation	Certification mark
Bureau Veritas Certification India Pvt. Ltd., Mumbai.	13-03-2010	NPOP; USDANOP	
ECOCERT India Pvt. Ltd., Aurangabad.	22-08-2011	NPOP; USDANOP	
IMO Control Pvt. Ltd., Bangalore.	27-09-2010	NPOP; USDANOP	
Indian Organic Certification Agency (INDOCERT), Cochin	24-10-2008 (Extended till next NAB meeting)	NPOP; USDANOP	
Lacon Quality Certification Pvt. Ltd., Thiruvalla (Kerala).	31-10-2011	NPOP; USDANOP	
Natural Organic Certification Agency (NOCA), Pune.	22-11-2009	NPOP; USDANOP	
OneCert Asia Agri Certification Pvt. Ltd., Jaipur.	26-10-2012	NPOP; USDANOP	
SGS India Pvt. Ltd., Gurgaon.	01-05-2011	NPOP; USDANOP	
Control Union Certifications, Mumbai.	28-05-2011	NPOP; USDANOP	

contd. ...

... contd.

Name of the certification agency	Validity of current accreditation	Scope of accreditation	Certification mark
Uttarakhand State Organic Certification Agency (USOCA), Dehradun.	13-11-2009	NPOP; USDANOP	
APOF Organic Certification Agency (AOCA), Bangalore.	09-01-2010	NPOP	
Rajasthan Organic Certification Agency (ROCA), Jaipur.	09-10-2010	NPOP	
Vedic Organic Certification Agency, Hyderabad.	30-09-2011	NPOP	
ISCOP (Indian Society for Certification of Organic Products), Coimbatore.	30-09-2011	NPOP	
Food Cert India Pvt. Ltd., Hyderabad.	30-09-2011	NPOP	
Aditi Organic Certifications Pvt. Ltd., Bangalore	30-09-2011	NPOP	
Chhattisgarh Certification Society, India (CGCERT), Raipur	15-09-2012	NPOP	
Tamil Nadu Organic Certification Department (TNOCD), Coimbatore	30-09-2012	NPOP	

(*Source*: Agricultural and Processed Food Products Export Development Authority - www.apeda.com as on 13.05.2010)

National Institute of Organic Farming—The Indian Central Government set up a National Institute of Organic Farming in October 2003 in Ghaziabad, Madhya Pradesh. The National Biofertilizer Development Centre was renamed as the National Institute of Organic Farming (NIOF). The purpose of this institute is to formulate rules, regulations and certification of organic farm products in conformity with international standards. Further details on current status of regulatory mechanism in organic farming in India may be obtained from Bhattacharyya and Gehlot (2004).

ORGANIC FARMING: CHARACTERISTICS AND ADVANTAGES

Characteristics of Organic Farming

Sustainable agriculture is necessary to attain the goal of sustainable development. "Sustainable agriculture is the successful management of resources for agriculture to satisfy changing human needs while maintaining or enhancing the quality of environment and conserving natural resources"- Food and Agriculture Organization (FAO). All definitions of sustainable agriculture lay great emphasis on maintaining an agricultural growth rate, which can meet the demand for food of all living things without draining the basic resources. Organic farming is one of the several approaches found to meet the objectives of sustainable agriculture. There are several definitions of organic farming and the one given by the United States Department of Agriculture (USDA) is considered the most coherent and stringent (Narayanan, 2005). Organic farming is defined as "a system that is designed and maintained to produce agricultural products by the use of methods and substances that maintain the integrity of organic agricultural products until they reach the consumer". This is accomplished by using substances, to fulfill any specific fluctuation within the system so as to maintain long term - USDA.

The definition given by IFOAM has also been referred as standard definition for organic agriculture. According to IFOAM, "Organic agriculture is a production system that sustains the health of soils, ecosystems and people. It relies on ecological processes, biodiversity and cycles adapted to local conditions, rather than the use of inputs with adverse effects. Organic agriculture combines tradition, innovation and science to benefit the shared environment and promote fair relationships and a good quality of life for all involved" (IFOAM). Unlike chemical farming, organic farming aims at "feeding the soil" rather than "feeding the plant". It is a means of giving back to the nature what has been taken from it. Since organic farming aims at maintaining soil health, the highest yields can be obtained in a sustainable and eco-friendly manner in the long run.

A large number of terms such as biological agriculture, ecological agriculture, bio-dynamic and natural agriculture are used as an alternative to organic farming. Organic farming is characterized by the maximal, but sustainable use of local resources and minimal use of purchased inputs. It facilitates the basic biological functions of soil-water-nutrient-human continuum. In organic farming the ecological balance is maintained as the bio diversity is being sustained. Organic farming is a form of agriculture that relies on crop rotation, green manure, compost, biological pest control, and mechanical cultivation to maintain soil productivity and control pests, excluding or strictly limiting the use of synthetic fertilizers, chemical pesticides, plant growth regulators, livestock feed additives, and genetically modified organisms. Enhancing soil health is the cornerstone of organic farming.

Though the organic farms produce lower yields, they require no synthetic fertilizers and pesticides. The decreased costs on those inputs, along with the premiums, which consumers pay for organic produce, create higher profits to organic farmers. Moreover, it has already been realized in certain areas that the yields in organic farming are far greater than those obtained by chemical farming in the long run. There is often a loss in yield while converting a chemical farm in to organic farm, but after few years the yield will certainly increase and get stabilised at a level often higher than that achieved under a chemical regime. It is therefore recommended to convert gradually over a period of three to four years, if income from the farm is a key issue (www.satavic.org).

Indigenous seeds, crop rotation, intercropping, multicropping, biological control of pests and diseases, botanicals pesticides, reduced tillage, mulching, green manuring and green leaf manuring, composting, vermicomposting, biofertilizers, soil and water conservation are some of the techniques and practices integral to organic farming. This list is not complete in itself, and farmers can innovate and adapt to suit their surroundings.

Advantages of organic farming

The various benefits of organic farming all over the world include high premium, low capital investment and the ability to use traditional knowledge. Given below are some of the advantages of organic farming for small farmers in India (www.organicfacts.net):

- ***High premium:*** Organic food is normally priced 20-30% higher than conventional food. This premium is very important for a small farmer whose income is just sufficient to feed his/her family with one meal.
- ***Low investment:*** Organic farming normally does not involve capital investment as high as that required in chemical farming. Further, since organic fertilizers and pesticides can be produced locally, the yearly costs incurred by the farmer are also low. It should also be noted that while shifting from chemical farming to organic farming, the transition might be costlier.
- ***Traditional knowledge:*** Small farmers have abundance of traditional knowledge with them and within their community. Most of this traditional knowledge cannot be used for chemical farming. However, when it comes to organic farming, the farmers can make use of the traditional knowledge. Further, in case of organic farming, small farmers are not dependent on those who provide chemical know-how.

In India more than 80% of the farmers are small and marginal farmers and many small farmers are practicing organic farming; since they are unaware of the market opportunities, they are not able to reap the benefits of organic farming mentioned above. Hence necessary steps need to be taken to address this issue.

CONSTRAINTS IN ORGANIC FARMING

Though there are several benefits in adapting the organic farming, certain barriers hinder the adaption at farmers' level. Although the land resources can move freely from organic farming to conventional farming, they do not move freely in the reverse direction. Initial crop loss is one of the major constraints while transforming the land from chemical to organic agriculture. The biocontrol agents may take three to four years to build up in a crop ecosystem. Moreover, their rate of success is in general, more in horticultural eco system but, in annual crop ecosystem their establishment is always questionable. Above all, farmers may afraid to adopt the organic agriculture without adequate support from the government and premium price for their produce. Thus, organic agriculture has its own limitations. These limitations were studied by the Office of Evaluation and Studies (OE), International Fund for Agriculture Development (IFAD). They include the following:

- Organic farming is labour intensive. Hence it is beneficial to a small farmer who has abundant labour in his/her family. A small farmer who does not have sufficient labour at home may find it difficult to hire external labour and therefore organic farming may not be feasible for him/her.
- Organic farming may also turn out to be expensive given the situation when a small farmer has to carry out the transition, modify the soil structure drastically or get a certification.
- Though a farmer can use a good deal of traditional knowledge in organic farming, since the practice is not as common as chemical farming, it is difficult for good scientific organic farming practices to propagate to the small farmers.
- Since organic produce is not traded in many markets, marketing the organic produce may be difficult for a small farmer.
- In organic farming, soil fertility needs to be replenished with the application of organic fertilizers regularly. If the organic fertilizers are not produced locally, it might be difficult for the small farmer, who would purchase the fertilizers in small quantities to obtain them from elsewhere. Further due to low economies of scale, there might be a huge transportation cost associated with the organic fertilizers obtained from long distances.
- It has also been observed that organic food prices are not stable and keep fluctuating from time to time. In such a scenario a low price might affect a small farmer drastically.
- Since the volume of organic food production is low (about 1-2% of total food production), it is very difficult to implement social security measures such as minimum support price.
- Certification is also an important aspect of organic farming. In many cases, providing education related to necessary certifications and the guidelines that need to be followed for obtaining the certifications may be difficult. Since certification cost is exorbitantly high and inorganic input use is prohibited in the nearby areas, organic farming may not be so easy to practise by the small and marginal farmers who constitute about 80% of the Indian farming community (Fertilizer Association of India, 2004).

ORGANIC FARMING IN INDIA: THE CONTROVERSIES

There are three categories of opinions about the relevance of organic farming in India. The first one simply dismisses it as a fad or craze. The second category, which includes many farmers and scientists, opines that there are merits in the organic farming, but we should proceed cautiously considering the national needs and conditions in which Indian agriculture functions. They are fully aware of the environmental problems created by the conventional farming. But many of them believe that yields are lower in organic cultivation during the initial period and also the cost of labour tends to increase therein. The third one is all for organic farming and advocates its adoption wholeheartedly. They think that tomorrow's ecology is more important than today's conventional farm benefits. However, among many a major reservation, the profitability of organic farming vis a vis conventional farming is the crucial one from the point of view of the Indian farmers, particularly the small and marginal (Narayanan, 2005).

Organic farming and its implications on India's food security were well discussed by Chhonkar and Dwivedi (2004). Some of the salient issues are presented hereunder:

- The widely publicised advantage of high income through export of organic produce to the western markets appears lucrative, but not so easy in real sense. Indigenous organic food market is in infancy and may remain so in the future. There is a remote possibility of a resource-poor small farmer of a remote area getting premium price for his organic produce after successfully overcoming all the steps and formalities beginning from farm certification to the final export. The family demands of such farmers may compel them to ultimately dispose-off their organic produce in local markets without any premium.
- Chhonkar and Dwivedi (2004) warned that if a section of pro-change farmers of agriculturally important areas currently engaged in food grain production began to convert their farms to organic with the intention of getting premium price, there will be no option left for the country other than importing food grains as it has practiced a few decades back.
- Nitrate is the main end product of organic manure decomposition; moreover it is continuously released from the organic matter under decomposition. Since nitrate release is not synchronised with either crop demand or its uptake, it tends to accumulate in excess in soil and pose environmental risk. The ions irrespective of their origin whether organic or inorganic will behave similarly. There is no evidence that nitrate ions from organic sources are less mobile and have lower denitrification potential than from inorganic fertilizers. In addition, the trace elements and heavy metal concentrations present in the animal wastes and sewage sludges vary widely and at times very high and often exceed concentrations normally found in the inorganic fertilizers. Field application of these manures in large quantities may pollute the soil and there will be every chance to enter in to food chain and create health hazard.
- The biggest myth about the organic farming as indicated by Chhonkar and Dwivedi (2004) is that the country has enough organics available to replace chemical fertilizers to sustain present level of crop production. All tappable nutrients from organic sources will be barely able to meet the deficit of nutrients in soil after crop removal at present level of production.
- To maintain the present level of food grain production in India without input of inorganic fertilizers, the net additional area to be brought under cultivation will be more than the total geographical area of the country.
- There are contentions that organic farming is unsustainable. The comparatively lower yields and dependence of manure from low-yield cattle has prompted criticism that organic farming is environmentally unsound and incapable of feeding the world population. Among these critics are Norman Borlaug, father of the "green revolution," and winner of the Nobel Peace Prize, who asserts that organic farming practices can at most feed 4 billion people, after expanding cropland dramatically and destroying ecosystems in the process.
- A few case studies pertaining to organic production of wheat, cotton, tea and coffee in India were reported by Marwaha and Jat (2004). The overall economics of these initiatives was not in favour of organic farming. It was also reported

that in India, organic farming may be a myth, because of limited availability and restricted use of organic resources and ever growing need for food and feed.

INDIA'S EXPORT POTENTIAL OF ORGANIC PRODUCTS

Consumer demand for organic products is increasing across the globe. Global sales of organic food and drink have increased by 43% from 23 billion US-Dollars in 2002 to 33 billion US-Dollars in 2005 and 46 billion US-Dollars in 2007. Although all regions reported healthy increase, the two major engines driving growth are the European and North American markets. Global production of organic crops has not kept pace with demand with many sectors reporting undersupply. The North American market has been experiencing supply shortages for a number of years; however, Europe and Asia are now also affected. Thus, large volumes of imports are coming in from other regions (Sahota, 2007). The increasing demand for organic food products in the developed countries and the extensive support by the Indian government coupled with its focus on agri-exports are the drivers for the Indian organic food industry. The Indian government is committed towards encouraging organic food production. It allocated Rs. 100 crore during the Tenth Five Year Plan for promoting sustainable agriculture in India. With the domestic consumption being low, the prime market for Indian organic food industry lies in the US and Europe. APEDA (Agricultural and Processed Food Export Development Authority) coordinates the export of organic food in India (www.apeda.com).

India has now become a leading supplier of organic herbs, organic spices, organic basmati rice, etc. India produced 9,76,646 MT of certified organic products in 2007. India exported 86 items in 2007-08 with the total volume of 37,533 MT. The export realization was around US$ 100.4 million registering a 30% growth over the previous year. Organic products are mainly exported to European Union, USA, Australia, Canada, Japan, Switzerland, South Africa and Middle East. Cotton ranked first in terms of quantity of the products exported (16,503 MT). Cotton alone contributed 25% of the value of total organic products exported during 2007-08 followed by 13% by bamati rice (www.apeda.com as on 13.05.2010) (Table 7, 8). In February 2009, the export duty on export of Basmati rice has been withdrawn by the Government of India (Notification No. 10/2009-Customs) with a view to promote the organic products.

Table 7: Status of organic products in India (2007-08)

Total production	9,76,646 MT.
Total quantity exported	37,533 MT
Value of total export	Rs. 498 Crores (USD 100.4 million)
Total area under certification (including wild harvest)	2.8 million hectares
Total area under certified organic cultivation	0.45 million hectares
Share of exports to total production	4% (approx.)
Increase in export value over previous year	30% (approx.)

(*Source*: Agricultural and Processed Food Products Export Development Authority - www.apeda.com as on 13.05.2010)

Table 8 : Commodity wise export during 2007-08

Commodity	Export Contribution (of volume)	Export Contribution (Rs. In crores)	Export Contribution (of value)
Cotton	43%	123.88	25%
Basmati rice	15%	59.20	13%
Honey	11%	46.41	10%
Tea	8%	92.13	20%
Dry fruits	7%	84.31	18%
Processed food	5%	17.99	4%
Sesame	4%	9.13	2%
Spices	3%	20.09	4%
Medicinal & herbal plants/products	2%	10.59	2%
Others	2%	5.05	2%
Total		498.00	

(*Source*: Agricultural and Processed Food Products Export Development Authority - www.apeda.com as on 13.05.2010)

CONCLUSION

In India, only about 38% of the cultivated land is irrigated and in rest of the rainfed areas (62%) there is little or no use of fertilizers and other agrochemicals on account of associated risk in farming and poor resources. Therefore, there could be ample scope for pure organic farming in those areas. Thus, rather than promoting organic farming *en masse* and universally, it would be desirable to carefully delineate areas for organic agriculture, which provides better profitability and boost to the export (Marwaha and Jat, 2004). During the last four decades of the 20th century, the global population doubled from 3 to 6 billion and is estimated that by the year 2020, it will reach the 8 billion mark. Food and nutritional security is therefore a global concern. Neither conventional farming with inorganic alone nor the organic farming only with the use of organic inputs can face this challenge (Bhattacharya and Chakraborty, 2005). The combination of both organic and inorganic is undoubtedly the best option as on today for developing country like India. Above all organic agriculture must be discussed with an open mind, with advantages and disadvantages being clearly considered in order to create positive mindset among the decision makers, because the most important factor that will enable organic agriculture to usefully contribute to food security is the attitude of decision-makers (Wynen, 1998). India should be much more concern about organic agriculture than any other nation of the world as agriculture is the source of livelihood for more than 6 million people and it is the foundation for the economic development of the country.

SELECTED REFERENCES

Bhattacharya, P. and Chakraborty, G. 2005. Current status of organic farming in India and other countries. *Indian Journal of Fertilizers* **1**(9): 111-123.

Bhattacharya, P. and Gehlot, D., 2004. Current status of regulatory mechanism in organic farming. *Fertilizer News* **49**(11): 33-38.

Chhonkar, P.K. and Dwivedi, B.S. 2004. Organic farming and its implications on India's food security. *Fertilizer News* **49** (11): 15-31 & 38.

Herrmann, G.A. and Rundgren, G. 2007. Certification and Accreditation. *In*: The world of organic agriculture: statistics and emerging trends 2007 (Willer, H. and Yussefi-Menzler, M., Eds.). International Federation of Organic Agriculture Movements (IFOAM), Bonn, Germany and Research Institute of Organic Agriculture (FiBL), Frick, Switzerland, pp. 67-74.

Huber, B., Kilcher, L. and Schmid, O. 2007. Standards and Regulations. *In*: The world of organic agriculture: statistics and emerging trends 2007 (Willer, H. and Yussefi-Menzler, M., Eds.). International Federation of Organic Agriculture Movements (IFOAM), Bonn, Germany and Research Institute of Organic Agriculture (FiBL), Frick, Switzerland, pp. 56-66.

Kilcher , L. 2007. How organic agriculture contributes to sustainable development. *In*: The world of organic agriculture: statistics and emerging trends 2007 (Willer, H. and Yussefi-Menzler, M., Eds.). International Federation of Organic Agriculture Movements (IFOAM), Bonn, Germany and Research Institute of Organic Agriculture (FiBL), Frick, Switzerland, pp. 82-91.

Mahapatra, B.S., Ramasubramanian, T. and Chowdhury, H. 2009. Organic farming for sustainable agriculture: Global and Indian perspective, *Indian Journal of Agronomy*, 54(2): 178-185.

Marwaha, B.C. and Jat, S.L. 2004. Status and scope of organic farming in India. *Fertilizer News* **49**(11): 41-48.

Narayanan, S. 2005. Organic farming in India : relevance, problems and constraints. Occasional paper – 38. Department of Economic Analysis and Research, National Bank for Agriculture and Rural Development, Mumbai, 83p.

Sahota, A. 2007. Overview of the global market for organic food and drink. *In*: The world of organic agriculture: statistics and emerging trends 2007 (Willer, H. and Yussefi-Menzler, M., Eds.). International Federation of Organic Agriculture Movements (IFOAM), Bonn, Germany and Research Institute of Organic Agriculture (FiBL), Frick, Switzerland, pp. 52-55.

Stockdale, E.A., Lampkin, N.H., Hovi, M., Keatinge, R., Lennartsson, E.K.M., Macdonald, D.W., Padel, S., Tattersall, F.H., Wolfe, M.S. and Watson, C.A. 2001. Agronomic and environmental implications of organic farming system. *Advances in Agronomy* **70**:261-327.

Wynen, E. 1998. Evaluating the potential contribution of organic agriculture to sustainability goals. Environment and Natural Resources Service, Research, Extension and Training Division, Sustainable Development Department, Food and Agriculture Organization of United Nations (FAO), Rome.

Important websites consulted: www.organic-world.net; www.apeda.com; www.satavic.org; www.organicfacts.net.

❑❑❑

Chapter 3

Organic Farming in India : Perceptive and Prospects

KAPILA SHEKHAWAT* AND S.S. RATHORE**

*Scientist, Agronomy, Directorate of Rapeseed-Mustard Research Bharatpur.

**Senior Scientist, Agronomy, Directorate of Rapeseed-Mustard Research Bharatpur.

INTRODUCTION

The rising awareness towards organic agriculture is due to the biological magnification of pesticides and fertilizer residues in food chain and degradation soil natural fertility status. Injudicious use of inorganic nutrient sources and complete negligence for inclusion of organic sources as a part of fertility management has reduced the soil resilience to an ever time low level. Overuse of pesticides, especially in vegetables and fruits, resulted in residues above safety levels (Carson, 1963 and Agnihotri, 1999) and this brought to attention the ill-effects of modern agriculture; even drinking water was not spared. Soon the ill effects of overuse of fertilizer nitrogen were recognized worldwide and as well as in India. These were nitrate enrichment (nitrate bomb) of groundwaters; river waters and estuaries and volatilization of ammonia and nitrous oxide to the atmosphere; the former added to the problem of acid rain, while the latter led to the reduction of ozone layer (Datta *et al*, 1997). These ill-effects of modern agriculture forced people, especially in countries with high-income economies (European countries, USA, Canada, Australia, etc.) to demand food grown without fertilizers and pesticides. This paved the way for organic farming. Continued use of organic manure on a farm improves its organic matter content, which supports the soil micro, meso and macro fauna and makes the soil a living body (Ramakrishnan, *et al*, 2003).

The heath issues of toxic residues in food material are also a cause of worry to be continued with present practices of crop management in general and the management of soil fertility in particular. Other than the compulsions to go for organic farming, existence of some of the areas with no use or very low level of use of chemical has the additional advantages to adopt organic practices. There are many areas (like northeastern regions and arid and other rain fed areas) in our country where organic farming is being practiced with fully dependent on, on-farm natural resources . Due to these favorable conditions, a Task Force was constituted on Organic Farming under the chairmanship of Kunwarji Bhai Yadav. In his report the committee emphasized on the need for consolidating information on organic farming and its benefits. One of the steering committees constituted by this Task Force under the Chairmanship of M. S. Swaminathan, Chairman, National Commission on Farmers, Govt of India, has suggested taking up organic farming as a challenging task and as a thrust area of the 10th Five Year Plan. The steering committee advocated giving boost to organic farming in the northeastern states and in rainfed areas, where there is limited use of agricultural chemicals.

Organic farming is practiced in approximately 100 countries of the world and the area under organic management is continually growing. According to the SOEL-Survey (February 2004), more than 24 million hectares are managed organically world-wide. Currently, the major part of this area is located in Australia (about 10 million ha). The percentages of land under organic management, however, are highest in Europe. The total organic area in Asia is now about 0.88 Mha corresponding to 0.07 percent of the agricultural area. There is a wide spectrum of sector development stages, from early pioneer status to highly developed markets (Japan). The Japanese farmer and philosopher Masanobu Fukuoka practiced and promoted natural farming throughout the world and has followers in India also. His book "One-Straw Revolution" has been widely applauded. The formation of the International Federation of Organic Agriculture Movements (IFOAM) in 1972, gave an international framework for the discussion and codification of internationally recognized principles of organic farming. Organic farming as in the modern context was practiced in India only on 4800 ha in 2003 and the produce exported was valued at about 89 crores of rupees, which is only 0.8% of the current global market. Among the field crops, only Basmati rice, cotton and sesame were exported (Rajendra Prasad, 2005).

To the maximum extent feasible, Organic farming system rely upon crop rotations, crop residues, animal manures, off-farm organic wastes, mechanical cultivations, mineral bearing rocks, and aspects of biological pest control to maintain soil productivity, to supply plant nutrients and to control insects, weeds and other pests.

USDA has defined organic farming as "It is a production system which avoids or largely excludes the use of synthetically compounded fertilizers, pesticides, growth regulators and livestock additives." Codex Alimentarius Commission, a joint body of FAO/WHO defines "organic agriculture as holistic food production management systems, which promotes and enhances agro-ecosystem health, including biodiversity, biological cycles and soil biological activity. It emphasizes the use of management practices in preference to the use of off-farm inputs, taking into account that regional conditions require locally adapted systems. This is accomplished by using, where possible, agronomic, biological and mechanical methods, as opposed to using synthetic materials to fulfil any specific function within the system.

SCENARIO OF ORGANIC FARMING TRADE IN INDIA

Various groups have estimated the area under organic agriculture in India. The study undertaken by FIBL and ORG-MARG (Garibay and Jyoti, 2003) the area under organic agriculture to be 2,775 ha (0.0015% of gross cultivated area in India). Some of the major organically produced agricultural crops in India include crops like plantation, spices, pulses, fruits, vegetables and oil seeds etc. Almost 30.4 million ha land is managed organically by more than 700,000 farms world over. This constitutes 0.65% of the agricultural land of the countries covered under study. Oceania holds 42%, Europe 24%, Latin America 16%, Asia 11% and North America 7.2% Among Countries Australia is largest with 12.3 million ha, followed by China 2.3 mha, Argentina 2.2 mha and the US 1.6 mha (Soel and IFOAM survey, 2008).

Table 1 : Major product produced in India through organic farming

S.No.	Types of products	Products
1	Commodity	Tea, coffee, rice and wheat
2	Spices	Cardamon, black paper,ginger, turmeric, vanilla, tamarind, clove, cinnamon.
3	Pulses	Red gram and Black gram
4	Fruits	Mango, banana, pine apple, passion fruit, oranges, cashew nut, walnut
5	Vegetables	Okra, brinjal, garlic, onion, tomato, potato
6	Oilseeds	castor, mustard, sesamum, sunflower
7	Others	Cotton and herbal extracts

Source: Garibay and Jyoti, 2003

MAIN COMPONENTS OF ORGANIC CROP PRODUCTION

Biodiversity	Diversification and integration of components	Sustainability	Natural Plant Nutrition	Natural Pest management Diverse crop
• Crop Rotation • Green manure • Cover crop • Animal manure • Composting • Intercropping • Bio control • Farm scaping	Crop Rotation Animal Manures Composting Intercropping Mulching Farm scaping	• Crop Rotation Efficient cropping system • Cover crop • Intercropping • Bio control • Mulching • Resource conservation and efficient recycling	• Legumes in crop rotation • Composting • Cow dung slurry • Green manures • Vermicomposting • Farm Litter management • Crop residues management.	• Crop rotation • Green manure • Cover crop • Guard Crop • Break cropping • Use botanicals extracts • Weed and pest smothering crop. • Sanitation • Tillage • Biological traps

TECHNIQUES AND APPROACHES FOR REALISATION OF ORGANIC FARMING

There are some traditional practices for organic supplements and over the period many of refinements have already been made in these techniques. Also the Ministry of

Agriculture, Government of India has notified 4 bio-fertilizers and 3 organic fertilizers under Fertilizer Control Order.

Bio fertilizer

i. *Rhizobium*
ii. *Azotobacter*
iii. *Azospirillum*
iv. *P-solubilising biofertilizer*

Organic fertilizer

i. City Compost
ii. Vermicompost
iii. Press mud

Crop and crop residues management for nutrient supplement in Organic farming

i. Green manuring
ii. Live mulching
iii. Crop residues management

Legumes for restoration of soil fertility

Introduction of leguminous cover crops are gaining growing interest among farmers and researchers alike (Flores, 1992) for its multipurpose nature in enhancing soil productivity, weed suppression, providing food to humans and feeds to animals besides added cash incomes (Garrity and Lai, 2000). Velvet bean in Honduras is reported to contribute some 60 kg N/ha (Flores, 1992). He also reported that Brazilian researchers achieved maize yields of up to 6,800 kg/ha without any other external N-input but from the velvet bean alone. Similarly, Bunch (2001) reported that cover crops are capable of fixing more than 75 kg/ha of N, while velvet bean can fix even up to 140 kg/ha/crop.

Table 2 : Range of experimental estimates of the proportion (Pfix) and amount of N_2 fixed by important pulses and legume oilseeds

S.No.	Crop species	Pfix %	Amount of N fixed (kgN/ha)
1.	**Cool-season legumes**		
2.	Chickpea *(Cicer arietinum)*	8-82	3-141
3.	Lentil *(Lens culinaris)*	39-87	10-192
4.	Pea *(Pisum sativum)*	23-73	17-244
5.	Faba bean *(Vicia faba)*	64-92	53-330
6.	Lupin *(Lupinus angustifolius)*	29-97	32-288
7.	**Warm-season legumes**		
8.	Soybean *(Glycine max)*	0-95	0-450
9.	Groundnut *(Arachis hypogaea)*	22-92	37-206

contd. ...

... contd.

S.No.	Crop species	Pfix %	Amount of N fixed (kgN/ha)
10.	Common bean *(Phaseolus vulgaris)*	0-73	0-125
11.	Pigeon pea *(Cajanus cajan)*	10-81	7-235
12.	Green gram *(Vigna radiata)*	15-63	9-112
13.	Black gram (V. *mungo)*	37-98	21-140
14.	Cowpea (V. *unguiculata)*	32-89	9-201

Source: Bunch, 2001.

It is not only N which is fixed by legumes but some amount of P also fixed by legumes by release of P –solublising acids in the soils. The amount of N and P fixed by different leguminous crops has been depicted in the Table 2. After a legume crop matures its nodules disintegrate and release the bacteria they contain into the soil. These bacteria become a source of infection for subsequent legume crops, but the population decline gradually in the absence of suitable roots to nodulate and multiply on.

Earthworms for Building of Soil Fertility

The method of producing compost using earthworms is called vermin-composting. Earthworms constitute more than 80% of the invertebrate biomass. They play a significant role in maintaining the soil productivity. Earthworms help in decomposition of organic matter and contribute 20-100 Kg N per h per year apart from producing other mineralized nutrients and plant growth promoting substances. Earthworms help the growth of beneficial microorganisms by providing optimum conditions for their growth. Worms assimilate some of the microorganisms as their food. Hence, the more the microbial biomass the more will be the worm biomass. Worms enrich the soil with oxygen, thus encouraging aerobic microorganisms. These microorganisms perform several important functions like nitrogen fixation, production of enzymes, antibiotics, growth hormones, etc. They also control parasitic nematodes and enhance crop yields. Earthworms can feed on crop residues; vegetable and fruit wastes and convert them into Vermicompost. On an average one worm weighing about 1 g can produce about 1 g compost in a day. In an area of 1 sq. meter, if 1000 earthworms are inoculated, 1 kg of Vermicompost can be prepared in a day. This way, within 80-90 days the complete area of 1 sq. meter will turn into compost and the population of earthworms will increase several folds.

In traditional organic farming areas like in shifting cultivation of highly leached soils of the humid tropics, earthworm activity is particularly, important because of rapid incorporation of litter into the mineral soils and because of local concentrations of nutrients in the surface soil layers.

Soil Organic Carbon Management

Many tropical soils are poor in inorganic nutrients and rely on the recycling of nutrients from soil organic matter to maintain fertility. In undisturbed rainforests such nutrients are recycled via the litter; 'slash-and-burn' agriculture, meanwhile, depends on the mineralization of organic nutrients from the plant remains or on (short-lived) inputs from ash. The quantification of the role of organic matter in sustaining the fertility of soils under different climate zones will ensure good fertility status of soils. Tiessen, H, 1994 estimate rates of carbon turnover from ecological measurements and dating, and determine its relation to the soil carbon and nutrient budgets and reported that agriculture without supplementary fertilization was economical for 65 years on temperate prairie and for six years in a tropical semi-arid thorn forest.

Long-term experiments indicate that losses of up to 0.69 t carbon/ha/yr in the soil surface layers are common. The strategies required must include direct soil organic carbon (SOC) replenishment through addition of organic materials, notably manures, wastes, residues and plant litter; biomass transfer; incorporation of improved fallows in the farming systems. Indirect contribution to SOM replenishment or recapitalization may be achieved through inorganic fertilizations and amendments; legume integration in the production system; and combined inorganic and organic inputs (Nandwa, S.M. 2001).

Mulching as Source of Nutrients

Mulching is the agronomic practice of covering the soil surface either by live /dead plants or with other material like plastic sheets. The live mulch of leguminous herbs helps in nutrient enrichment of the soil. Hoang *et al*, 2002 examined the contribution of the leguminous bush *Tephrosia candida* (Roxb.) D.C. as a mulch producer to improve nutrient cycling and prevent nutrient losses by erosion. The *Tephrosia* mulching as fallow of *Tephrosia*, internal mulching using pruned *Tephrosia* biomass and upland rice with external mulching using *Tephrosia* biomass prevented nutrient losses by erosion effectively. This was a trend that the less labile P-pools (NaOH-P) were reallocated into the more labile P-pools (Bicarb-P) in the soil of the Burning released significant amounts of the inorganic P-pools. Sufficient quantity and quality of the *Tephrosia* mulching material was recorded with organic input to crop export ratios for N and P were >1.

Use of Charcoal

Charcoal is reported to be responsible for high soil organic matter contents and soil fertility. Higher nutrient retention and nutrient availability were found after charcoal additions to soil, related to higher exchange capacity, surface area and direct nutrient additions .Rapid turnover of organic matter leads to a low efficiency of organic fertilizers applied to increase and sequester C in soils of the humid tropics. Higher charring temperatures generally improved exchange properties and surface area of the charcoal. Additionally, charcoal is relatively recalcitrant and can therefore be used as a long-term sink for atmospheric CO_2. Several aspects of a charcoal management system remain unclear, such as the role of microorganisms in oxidizing charcoal surfaces and releasing nutrients and the possibilities to improve charcoal properties during production under field conditions. Several research needs were identified, such as field testing of charcoal production in tropical agro ecosystems, the investigation of surface properties of the carbonized materials in the soil environment, and the evaluation of the agronomic and economic effectiveness of soil management with charcoal (Glaser *et. al*, 2002).

The application of a mixture of manioc peel and charcoal, by improving crop production and soil fertility and enhancing earthworm activity, could be potentially efficient organic manure for legume production in tropical areas where manioc is cultivated under slash-and-burn shifting agriculture (Stéphanie Topoliantz *et. al*, 2005).

NON CHEMICAL WEED MANAGEMENT PRACTICES

Weeds are the major problem in crop production, there are perennial weeds which become noxious; to control them some of agronomic practices are very useful. As per one estimate weeds are the pest causing maximum loss in crop yield, which is up to 33 % and sometimes complete crop failures.

Management of Noxious Weed Thatch Grass (*Imperata cylindrica*)

Thatch grass (*Imperata cylindrica*) an obnoxious grassy weed normally establishes in patches in the agricultural land irrespective of altitude and agro-climatic zones. *Imperata* is known to be most troublesome weed especially in fallow land and short *jhum* cycles areas. Thatch is noxious weed especially in *jhum* field and once the field left as fallow thatch is coming up in big way. It is very difficult even to control by mechanical means due to its vigorous tillering and deep root system. Since its growth is fast and establishes soon in new land, control becomes very difficult and costly. A thatch infested *jhum* area is therefore, normally abandoned by the farmers after the first year cropping as it makes farm operations difficult and expensive to continue well into the second year. There are reports of control of thatch grass to a very large extent (about 50 %) by planting cassava in the first cropping season. Almost complete control was achieved by the end of the second cropping season at SARS, Nagaland. The grass initially colonizes in deep clay soil and later spreads extensively. It is very difficult to control or eradicate the grass once it establishes. However, from time to time farmers themselves have developed methods to manage this troublesome weed that are worth adapting and build upon their innovations. One such innovative method indigenous to the farmers is using of wild sunflower, which smothers the *Imperata* in just two years time.

Common Salt for Weed Control

The extremely acidic soils conditions in many of shifting cultivation areas helps in controlling the weeds though use of common salt. The modern commercial weedicides like 2,4-D and the conventional common salt are all Sodium (Na) based. Like common salt, application of Na salt of 2, 4-D can kill even sedges up to 90% besides increasing yield significantly. The farmers in north eastern region of India have been using common salt (Sodium Chloride) for controlling the broad-leaved weeds in their *jhum* fields. Common salt is probably the most effective herbicide (for broad-leaved weeds), while it does not effect the rice plant. However in subsequent cropping, the growth of narrow leaves increased more than normal in salt treated plots (NEPED, 2007).

SOME OF TRADITIONAL ORGANIC FARMING SYSTEM OF INDIA

Shifting Cultivation

Shifting cultivation, variously termed as rotational bush, fallow agriculture, swidden cultivation or slash and burn cultivation, is an age old practice of cultivation in north east part of India, which is still a predominant farming system.

The characteristics of *jhum* cultivation are as follows

(i) Slashing and clearing of forest areas and burning of the dried biomass by fire,

(ii) Rotation of *jhum* land every four to seven years,

(iii) Employ of human labour as the chief input,

(iv) Non-employment of animals implements or machinery,

(v) Collective ownership of land,

(vi) Reciprocal labour sharing and

(vii) Mixed cropping system.

Alder Based Farming System of North East India

The Alder-based *jhum* system is an outstanding model of sustainable land-use intensification evolved through numerous years of testing without the want of outside technologies. In North east India most of the farmers cultivate *jhum* fields for two years within a 10 years cycle (1:4 ratio of cropping to fallow), while Alder-based *jhum* system allows crop harvests in two out of every four years (1:1 ratio of cropping to fallow). Another unique advantage about alder system is that under short *jhum* cycles a number of crops can be produced in a sustainable way, and even produce saleable surplus items of food.

- Under-growths are slashed and cleared before pollarding. Alder trees are pollarded at a height of 2 meters from the ground (for initial pollarding).
- Before burn operation branches are collected for poles and fuel.
- Burning is done when twigs and leaves are completely dried, normally in February. Sowing is done in March-April.
- First year *jhum*, primary crop is rice. Along with rice number of secondary crops is grown. During second year, jobs tear and other secondary crops are grown. In one *jhum* field as many as 60 different crops were found.
- After two years of cropping, the fields are left fallow for another two years.

Alder-based *jhum* fields are managed typically in a four-year cycle, with two years of cropping between the alder trees fallowed by two more years while the soil is rested and the coppices are allowed to grow. *Alnus nepalensis* is a deciduous or semi-deciduous tree and it does not require high soil fertility but prefers permeable soils. It grows well on soils with high water content but not on waterlogged soils.

Alternate Land Use Systems for Organic Farming

The multi-disciplinary research programmes aimed at developing alternative land management practices has identified several viable land use models. For the conservation of natural resources and to save the farmers from misery there is dire need to diversify the cropping pattern by introducing alternative crops (Mittal, 2004).Watershed based farming system, appropriate soil conservation measures, mixed land use of agri-horti-silvi-pastoral system, subsidiary source of income through livestock rearing, creation of water harvesting and silt retention structure at lower reaches - these are the important distinguishing features of the suggested agricultural strategy . The model developed is based on the following distinct approaches:

Watershed, a natural drainage unit, should form the basis for planning various land uses to optimize the use of soil and water resources for sustained production. This watershed based farming system coupled with mechanical soil conservation measures viz. Contour trenches, contour bunds, bench terrace, half moon terrace, grassed waterway etc.

at appropriate locations can retain maximum rainfall within the slope, safely disposing off the excess runoff from the slopes to foot hills with non erosive velocity. Subsidiary income from rearing of livestock by feeding on the by-products of crops and cultivated fodders, trees raised on the terrace risers, bund surface and very steep slopes unfit for cultivation.

Agro Forestry System

Agro forestry refers to land use systems in which trees or shrubs are grown in association with crops (agricultural crops or pastures), in a special arrangement or a rotation; and in which there are both ecological and economic interactions between the trees and components of the system (Young, 1989).

COST CONSIDERATION OF ORGANIC FARMING

Monetary and Non Monetary Inputs

To practice organic farming, a sound package of operation is required to be followed in holistic manner. During initial period of conversion from intensive modern farming to organic farming a decline in the productivity has been reported by many workers. Complete reliance on organic inputs right from the supply of plant nutrients and insect pest, disease and weed management, the cost of cultivation is escalating as compare to other form of cultivation. As for input supply organic farming is complete dependent on, on-farm resources, if managed efficiently and in eco-friendly manner, there is enough chance to reduce the cost of cultivation. The efficient, effective and complete recycling of farm wastes will help not only in reducing the cost but in maintaining the eco balance and microclimate in most acquiescent way. There is need of twin pillar strategy to enhance the total factor productivity as genetic enhancement and improved agro techniques (CGIAR, 1998 and Paroda, 1998) .The main components of agro techniques of any cultivation, generally and organic farming, especially are monetary and non monetary. Monetary components encompasses technical inputs, fixed and variable cost components and among non monetary inputs includes certain agro techniques which do not require additional monetary resources but have the great impact on growth, yield attributes and yield by that way improve the total income of the farmers.

Certified organic products are normally more expensive than their conventional counterparts for a number of reasons :(FAO)

- Organic food supply is limited as compared to demand;
- Production costs for organic foods are typically higher because of greater labour inputs per unit of output and because greater diversity of enterprises means economies of scale cannot be achieved;
- Post-harvest handling of relatively small quantities of organic foods results in higher costs because of the mandatory segregation of organic and conventional produce, especially for processing and transportation;
- Marketing and the distribution chain for organic products is relatively inefficient and costs are higher because of relatively small volumes.
- As demand for organic food and products is increasing, technological innovations and economies of scale should reduce costs of production, processing, distribution and marketing for organic produce.

Non Monetary Inputs in Organic Crop Production for Reducing the Cost of Organic Cultivation

There are always concern to reduce the cost of cultivation of organic farming to make it further more economically viable on the one hand and affordable to common man on the other hand. In today's globalise era it is essential to make the agricultural produce cost effective to remain in the international market. The organically produced commodities are having potential international market but to be successful in the international market the cost consideration are utmost important. In most of the efforts for reduction of cost of cultivation , it is mainly monetary inputs discussed at all level including policy, planning, academic and societal level. Monetary components as encompass technical inputs, fixed and variable cost components, on the other hand the non monetary inputs includes certain agro techniques which do not require additional monetary resources but have the great impact on growth, yield attributes and yield by that way improve the total income of the farmers are not getting the deserved attention. It has been proved by many workers that the non monetary inputs have the great significance in boosting the crop productivity and simultaneously in reduction of cost of production.

The practices come under nonmonetary components are generally given less attention because of many reasons as lack of awareness about their impact, lack of proper demonstration, training to the farmers, lack of relevant literature etc. The non monetary practices generally considered can be enumerated as below-

- timely sowing,
- line-sowing,
- Crop geometry
- maintaining optimum plant population,
- timely irrigation at critical stages of crop growth
- Contingency crop planning under aberrant weather conditions
- Consideration of critical period competition in weed management
- and efficient use of bio-fertilizers,
- efficient water management
- organic manures, green manuring to maintain soil health
- Plant protection measures.
- Sound integration of on farm resources for efferent recycling of the wastes.

CONCLUSION

Organic farming holds great promises in our country due to growing concern to the health and environment issues. Organic farming relies on ecological processes, biodiversity and cycles adapted to local conditions, rather than the use of inputs with adverse effects. Organic agriculture combines tradition, innovation and science to benefit the shared environment and promote fair relationships and a good quality of life for all involved. Agro-climatic conditions in India and our agricultural biodiversity are favourable to organic agriculture and hence, offer tremendous scope for cultivation of a wide range of organic products. Keeping in foreground the food security of the nation, organic farming can be encouraged as practice in selected areas where use of technical inputs as agricultural chemicals is already very low. Most of India's farms are "organic by default." These areas are north-eastern region and rain fed area of the country. In India the organic agriculture is receiving increasing attention among farmers, producers, traders, exporters and consumers. Organic agriculture is a production system that sustains the health of soils, ecosystems and people. In India, the practice of organic farming in selected areas has already been started and leading to sustain high agricultural productivity.

REFERENCES

Agnihotri, NP 1999. Division of Agricultural Chemicals, Indian Agricultural Research Pesticide: Safety Evaluation and Monitoring. All India Co-ordinated Project (AICRP) on Pesticide.

Bunch, R., 2001. A proven technology for intensifying shifting agriculture: green manure / cover crop experience around the world. IIRR Resource Book. pp. 210-217.

Carson, R. 1963. Silent Springs, Hamish Hamilton, London, 1963.

CGIAR System Review Secretariat. 1998. Third system Review of The Consultative Group of International Agricultural Research (CGIAR). International Research partnership for Food Security and Sustainable Agriculture. CGIAR, Washington DC.

Datta, P. S., Deb, D. L. and Tyagi, S. K., J. 1997. *Contam. Hydrol.*, 27: 249–262.

Flores S.M. 1992. Practical use of cover crops. *ILEA Newsletter* 8(4): 15.

Food and Agriculture Organization of the United Nations (FAO) 2002. Organic agriculture, environment and food security, *Environment and Natural Resources*, 252.

Fukuoka, M. 1978. One-straw Revolution, Rodale Press, USA,.

Garibay S V and Jyoti K, 2003. Market Opportunities and Challenges for Indian Organic Products, Study funded by Swiss State Secretariat of Economic Affairs.

Garrity, D.P. and Lai, C.K., 2000. Innovation in shifting cultivation in Asia: Indigenous Fallow Management. *ILEA Newsletter*, 16(3): 5-6.

Glaser,-B.; Lehmann,-J. and Zech,-W. 2002. Ameliorating physical and chemical properties of highly weathered soils in the tropics with charcoal. Biol-fertil-soils. Berlin, Germany: Springer-Verlag. 35(4): 219-230.

Mittal. S.P. 2004. (edited) Alternative Land Use Systems for Sustainable Production/ Dehra Dun, Oriental Enterprises, 2004, xiv, 493 p.

NEPED 2007. Adding value to shifting cultivation in Nagaland, India. Nagaland Empowerment of People through Economic Development under International Development Research Centre and India- Canada Environment Facility. Nagaland, India. pp 211.

Nandwa,-S.M. 2001. Soil organic carbon (SOC) management for sustainable productivity of cropping and agro-forestry systems in Eastern and Southern Africa. Nutr-cycl-Agroecosyst. 61 (1/2): 143-158.

Paroda R.S. 1998. Paper published in souvenir of first International Agronomy Congress, October, 1998 at IARI, Pusa, and New Delhi. Project on Residues, IARI, New Delhi, 1999.

Rajendra Prasad 2005. Organic farming vis-à-vis modern agriculture. Current Science, 89(2): 252-254.

Ramakrishnan, P.S., Swift, M.J., Saxena, K.G., Rao, K.S., and Maikhuri, R.K. 2003. Soil Biodiversity, Ecological Processes and Landscape Management. Oxford & IBH, New Delhi.

Technical Report 2001. Development of sustainable land use options for shifting cultivators in Nagaland. State Agricultural Research Station Yisemyong::Mokokchung.

Stéphanie Topoliantz, Jean-François Ponge and Sylvain Ballof. 2005. Manioc peel and charcoal: a potential organic amendment for sustainable soil fertility in the tropics Biology and fertility of soils 41(1): 15-21.

Tiessen,-H. 1994. The role of soil organic matter in sustaining soil fertility. Nature. London : Macmillan Magazines Ltd. Oct 27,. v. 371 (6500) p. 783-785.

Young, A. 1989. Agroforestry for soil conservation. Wallingford, UK, C.A.B. International/ ICRAF. 276 pp.

❑❑❑

Chapter 4

Standards, Certification and Inspection in Organic Agriculture

LIJO THOMAS AND SETHULEKSHMI K.R.

Scientist, Agricultural Economics, Directorate of Rapeseed Mustard Research, Bharatpur, Rajasthan

Organic agriculture has developed over the years from a generic and broadly defined term with many interpretations to a well defined concept with concrete boundaries and limits. FAO defines organic agriculture as a unique production management system which promotes and enhances agro-ecosystem health, including biodiversity, biological cycles and soil biological activity, and this is accomplished by using on-farm agronomic, biological and mechanical methods in exclusion of all synthetic off-farm inputs. The undocumented traditional production systems which existed have given way to well documented production , processing and marketing systems with the development of commercial interest in organic farming . The implications of these changes have been far reaching in nature. A class of economic agents including producers, consumers and other intermediaries have emerged, who demand ever increasing accuracy and information regarding the goods and services being offered in the market.

BACKGROUND OF ORGANIC STANDARDS AND CERTIFICATION

Long before any governmental regulation of organic farming with respect to standards and certification came into existence, private standards were being followed, which were developed by individual farmers, farmer groups and consumer associations. These were very rudimentary in nature and were of unsophisticated. Nevertheless, they served the purpose for which they were developed which is evident from the fact that these standards evolved very quickly due to their perceived utility. Most of these groups invited voluntary participation from the farmer producers in the locality to become members of the group and to abide by the agreed standards of the group.

The first international standard was developed by the private, non-profit organization, called the International Federation of Organic Agricultural Movements (IFOAM). This set of basic standards synthesized a consensus on the basic parameters for what organic agriculture stands for and created a common ground for further development in the field of standards for organic agriculture. The IFOAM standards gave adequate flexibility which implicitly recognized the role of local biological and non biological factors, while developing the standards, that there would always be local adaptation. The European regulation was one of the first comprehensive set of standards developed by public regulatory system. The European regulations have , since then significantly influenced the standards developed by government bodies in many other countries. The influence of either of the three major importers of organic produce in the world, viz, European Union, USA and Japan can be found in the standards adopted by most of the other countries. This serves a practical purpose of being in congruence with the required export quality demanded by these countries. The Codex Guidelines (1999), which is now considered to be a major multilaterally accepted standard for organic production can also seen to be influenced by the EU regulations and IFOAM standards. Initially the focus of the standards and regulations were the products themselves and the production standards. These standards defined what the farmer or the producer had to do to produce organic products. With the emergence of multiple players in organic production , marketing and certification , the focus shifted towards developing standards and regulation on how the process of certification is being done and the requirements of certification and inspection agencies. There exist a vast number of standards and certification criteria, followed by different countries and organizations. The essential content of most of them remains the same, though they differ in certain aspects. The quest for an universally acceptable standard may not only be futile but also an imprudent one.

CONCEPTS

Standards

Standards are essentially the document approved by a recognized body, containing rules, guidelines or characteristics for products, processes and production methods for organic produce. Compliance to these standards is not mandatory, but voluntary in nature. The standards often include definitions of terminologies related to organic farming. It may also include information regarding symbols, packaging, marking or labelling requirements applicable to product, process and production methods. The standards usually serve as a guide post for common and repeated use by certifying agencies in the process of certification. Organic standards describe minimum requirements for a farm or product to be certified as 'organic'. There are organic standards on the national as well as international level developed by public and private agencies. Globally there are more than 60 standards which include IFOAM basic standards, CODEX Alimentations Commission guidelines EU Regulations and the NOP of USA. Though there is considerable degree of similarity between standards developed by different agencies at the international level, there exist certain differences between the various standards. The national standards and private standards have strived to adapt the basic features of the international standards like IFOAM standards and Codex standards to the local requirements and conditions. Standards have been developed for specific products and for specific purposes. Most organic standards cover the basic activities of crop production, livestock production, wild harvest or wild collection, handling and processing and labeling.

Certification

The procedure by which a third party gives written assurance that a clearly identified process has been methodically assessed, such that adequate confidence is provided that specified products conform to specified requirements. Certification is primarily a market mechanism to enable producers or traders of a product to access specific markets. These markets require that the products confirm to specified quality parameters. The certification process often results in the product fetching a premium price. The output of certification is the issue of a certificate and the right to use an approved commonly recognized mark or symbol which identifies the product as one processing specific qualities. Certification is usually performed against a standard or technical regulation. The system of production is verified against an organic standard. The commonly followed practice is the certification through assessment by an accredited third party. There are nearly 500 certification agencies globally. These certification bodies have been accredited to do the certification work by different accreditation authorities. The certification agencies are expected to develop and operate based on certain good practices as elaborated under ISO/IEC /IEC 65, the international standard for certifying agencies. The good practices include Non discrimination, Impartiality, Responsibility, Objectivity, Participation, Quality improvement, Confidential and Transparency. The cultivated land under certification is around 2.8 million ha (2007-08). This includes 1 million ha under cultivation and the rest is under forest area (wild collection)

The certification agencies are distributed unevenly across the globe with developed regions like European Union and developed countries like USA, and Canada accounting for the bulk of the certification agencies. Lately, the developing countries have also started to develop institutional capacity and capability for conducting certification and inspections. In India several certification agencies are in operation , catering to the needs of producers, processors and traders of organic products. Some of these certification agencies are international firms having operations in India (eg:- ECOCERT International, LACON GmbH etc.). Certification agencies entirely based in India are also in operation. (eg:- INDOCERT)

Accreditation

Accreditation is the procedure by which an authoritative body, usually a government agency, gives a formal recognition that a body or person is competent to carry out specific tasks. The process of certification is generally done through agencies accredited by the authoritative body. Accreditation is just another layer of assurance. The accreditation body provides an assurance of competence regarding the certification body which gives assurance regarding the product. The process of accreditation of certification agencies harmonises the entire process of production and processing of organic produce in different parts of the area under its jurisdiction. Accreditation facilitates trade as it allows the authorities, the traders and consumers to accept the assurance of certification bodies that may be operating on different parts of the world. Accreditation improves the quality of the certification process and reduces the risk of faulty and fraudulent certification, thereby protecting the interest of different stakeholders. It helps to create uniformity in the production process of organic products.

The international standard for accreditation bodies is ISO/IEC /IEC 17011. This lays down the good practices to be followed by accreditation bodies. In India, accreditation is done by the National Accreditation Body. Presently, the National Steering Committee for National Programme on Organic Production acts as The National Accreditation Body. The National Programme on Organic Production document states that The National Steering Committee may amend the Accreditation procedures whenever it deems fit. The National Accreditation Policy and Programme is subject to periodic internal review, which will be conducted by the Technical Committee, which will advise the National Steering Committee about the need and content of such amendments in the National Programme for Organic Production.

NEED AND SIGNIFICANCE OF STANDARDS AND CERTIFICATION

There are several benefits for stakeholders of organic agriculture in developing standards and certification mechanism.In many cases the organically and non- organically produced products get mixed up during the process of marketing and distribution. The end result is that by the time the product reaches the consumers, it becomes very difficult to identify them separately. In such cases, it becomes difficult to develop a separate market for organically produced product hurting the interest of the producers of organic products. This situation exists in many parts of Indian sub continent. The solution for this situation lies in creation of a separate organic market through certification. The certification mechanism also provides the following benefits

- Protects the organic producers from unfair competition
- Provides confidence to the consumers regarding the quality of the produce
- Facilitates international trade through use of common certification standards
- Develops and creates quality consciousness among the producers and consumers
- Helps the producers to realise premium price for their produce

Organic certification confirms that a product has been grown and processed according to specific organic standards. Labels and certification help the consumers to recognize trustworthy organic products easily and to build consumer confidence.

INTERNATIONAL STANDARDS

As mentioned earlier there are organic standards on the national as well as international level developed by public and private agencies. Here two of the most comprehensive international standards are introduced briefly, one developed by the International Federation of Organic Agriculture Movements (IFOAM), an international umbrella organization for organic organizations established in 1972 and the second by the inter-governmental body Codex Alimentarius Commission. IFOAM first published their so-called 'Basic Standard' in 1980 and it has been updated and amended every few years since. The standards are described as being 'standards for standards' which means they are not intended to be used by operators but for providing a framework for standard setters for their own locally adapted standards. According to IFOAM the Basic standards developed by them covers the following areas

- Organic ecosystems – including conservation, GMOs and wild harvest,

- General requirements – including conversion and split production
- Crop production
- Livestock including bees
- Aquaculture
- Processing and handling
- Textile processing
- Labelling
- Social Justice

The IFOAM Criteria, together with the Basic Standard, form the IFOAM Norms, which as a whole are the basis for the IFOAM Accreditation Programme for certification bodies.

The Codex 'Guideline for the production, processing, labelling and marketing of organically produced foods' was first published in 1999 and amended in 2001. The Codex Guideline has a different structure to the IFOAM standard but covers similar topics. The Codex Alimentarius Commission guidelines stated in their foreword their aims as:

- To protect consumers against deception and fraud in the market place and unsubstantiated product claims;
- To protect producers of organic produce against misrepresentation of other agricultural produce as being organic;
- To ensure that all stages of production, preparation, storage, transport and marketing are subject to inspection and comply with these guidelines;
- To harmonize provisions for the production, certification, identification and labelling of organically grown produce;
- To provide international guidelines for organic food control systems in order to facilitate recognition of national systems as equivalent for the purposes of imports; and
- To maintain and enhance organic agricultural systems in each country so as to contribute to local and global preservation.

Both these standards envisage the requirement for local adaptation and have provided for scope of improvisation within the standards. This has enabled the agencies vested with the duty to develop National standards in different countries to adopt and adapt these international standards substantially.

INSTITUTIONAL STRUCTURE FOR ORGANIC CERTIFICATION IN INDIA

The requirement of a strong institutional mechanism for organic certification was being increasing felt with the growing market base and prominence of organic farming in India. The lack of common comparable standards for certification and lack of planning and coordination in the process of certification lead to the slow growth of the market for organic produce. Moreover, it was felt that to build the customer confidence in organic products, a strong certification institutional mechanism is required. To provide a focused and well directed development of organic agriculture and quality products, Ministry of commerce and Industry, government of India, launched the National Programme on Organic Production (NPOP) in the year 2000. NPOP was formally notified in 2001 under the Foreign Trade and Development Act.

The NPOP document provides information on standards for organic production, systems criteria and procedures for accreditation of inspection and certification bodies, the national organic logo and the regulations regarding its use. The National Programme for Organic Production shall, among others, include the following:

(a) Policies for development and certification of organic products
(b) National standards for organic products and processes
(c) Accreditation of programmes operated by Inspection and Certification Agencies
(d) Certification of organic products

The NPOP will be developed and implemented by the Government of India through its Ministry of Commerce and Industry as the apex body. A National Steering Committee for National Programme for Organic Production has been constituted to oversee the implementation of NPOP. The members of the steering committee shall be drawn from Ministry of Commerce and Industry, Ministry of Agriculture, Agricultural and Processed Food Products Export Development Authority (APEDA), Coffee Board, Spices Board and Tea Board and other government and private organisations associated with the organic movement. The National Steering Committee will have sub committees to advice the committee on issues pertaining to national standards and accreditation. The National Steering Committee for National Programme on Organic production have drawn up the National Standards for Organic Products, which includes standards for organic production and processes as well as the regulations for use of the National Organic Certification Mark. For accreditation purpose, a National Accreditation Body has been created under the NPOP. This Accreditation Body is vested with the duty to administer the National Accreditation Policy and Programme laid down by The National Steering Committee on organic production. The National Steering Committee on organic production itself functions as the National Accreditation Body. The members of the National Accreditation Body shall comprise of representatives from Ministry of Agriculture, Ministry of Commerce and Industry, APEDA, Coffee Board, Spices Board and Tea Board. The Chairman of the Body shall be the Chairman of the National Steering Committee. National Accreditation Policy and Programme is subject to periodic internal review, which will be conducted by the Technical Committee. This committee will advise the National Steering Committee about the need and content of such amendments in the National Programme for Organic Production. According to the NPOP the work of the National Accreditation Body shall include;

1. Drawing up procedures for evaluation and Accreditation of certification programmes.
2. Formulating procedures for evaluation of agencies implementing the programmes.
3. Accreditation of inspection and certification agencies

An Evaluation Committee, comprising of members drawn from various government agencies and appointed by the National Accreditation Body, shall evaluate eligible Inspection and Certification agencies. APEDA, on behalf of the National Accreditation Body, will receive and screen applications from the certification agencies, will coordinate and arrange evaluation visits etc. to ascertain the credentials of certification programmes of the applicants. The Evaluation Committee will submit its recommendations to the National Accreditation Body for considering accreditation.

Operational Structure of National Programme for Organic Production (NPOP)

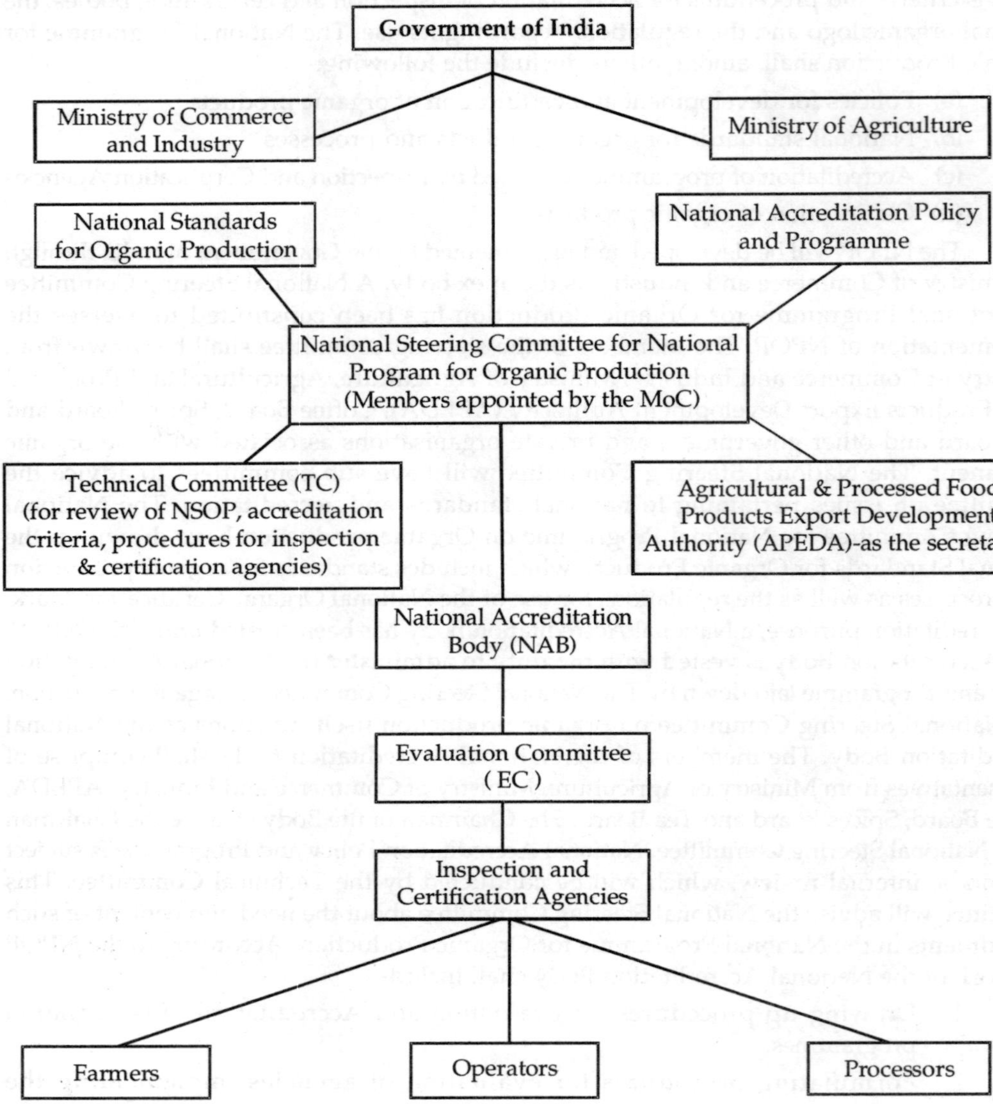

Source: NPOP Document , Ministry of commerce and Industry, GoI

NATIONAL STANDARDS FOR ORGANIC PRODUCTS IN INDIA (NSOP)

In 2000, the Government of India released the National Standards for Organic Products (NSOP) under the National Programme for Organic Production (NPOP). Products sold or labelled as 'organic' need to be inspected and certified by a nationally accredited certification body. The standards and procedures have been formulated in harmony with the international standards like the codex and IFOAM standards and keeping the specific requirements of our country in mind. Inputs provided by different agencies like government departments, commodity boards, producers cooperatives and NGOs have been given due consideration while developing the national standards for Organic Farming.

These NSOP standards are grouped under following six categories:

1. General requirements for Conversion
2. Crop production
3. Animal husbandry
4. Food processing and handling
5. Labeling
6. Storage and transport

The salient features of the standards elaborated in NSOP under General requirements for conversion, crop production and food processing & handling, which are highly relevant for agriculture sector are given below.

A. General Requirements

Conversion

- The standards requirements shall be met during the conversion period. All the standards requirements shall be applied on the relevant aspects from the beginning of the conversion period itself.
- If the whole farm is not converted, the certification programme shall ensure that the organic and conventional parts of the farm are separate and inspectable.
- Before products from a farm/project can be certified as organic, inspection shall have been carried out during the conversion period.
- Simultaneous production of conventional, organic in conversion and/or organic crops or animal products which cannot be clearly distinguished from each other, will not be allowed.
- To ensure a clear separation between organic and conventional production, a buffer zone or a natural barrier should be maintained.
- A full conversion period is not required where de facto full standards requirements have been met for several years and where this can be verified through several means and sources

Maintenance of Organic Management

- Converted land and animals shall not get switched back and forth between organic and conventional management.

Landscape

- The certification programme shall develop landscape and biodiversity standards.

B. Crop Production

Choice of Crops and Varieties

- When organic seed and plant materials are available, they shall be used. The certification programme shall set time limits for the requirement of certified organic seed and other plant material.

- When certified organic seed and plant materials are not available, chemically untreated conventional materials shall be used.
- The use of genetically engineered seeds, pollen, transgene plants or plant material is not allowed.

Duration of Conversion Period

- Plant products can be certified organic when the national standard requirements have been met during a conversion period of at least two years before sowing or in the case of perennial crops other than grässland, at least three years before the first harvest of products.
- The conversion period can be extended by the certification programme depending on, e.g., past use of the land and environmental conditions.
- The certification programme may allow plant products to be sold as "produce of organic agriculture in process of conversion" or a similar description during the conversion period of the farm.
- For the calculation of inputs for feeding, the feed produced on the farm unit during the first year of organic management, may be classified as organic. Feed produced on the farms in accordance with the national standards is to be preferred over conventionally grown / brought-in feeds.

Diversity in Crop Production

- Where appropriate, the certification programme shall require that sufficient diversity is obtained in time or place in a manner that takes into account pressure from insects, weeds, diseases and other pests, while maintaining or increasing soil, organic matter, fertility, microbial activity and general soil health. For non perennial crops, this is normally, but not exclusively, achieved by means of crop rotation.

Fertilisation Policy

- Biodegradable material of microbial, plant or animal origin shall form the basis of the fertilisation programme.
- The certification programme shall set limitations to the total amount of biodegradable material of microbial, plant or animal origin brought onto the farm unit, taking into account local conditions and the specific nature of the crops.
- The certification programme shall set standards which prevent animal runs from becoming overmanured where there is a risk of pollution.
- Manures containing human excreta (faeces and urine) shall not be used.
- Mineral fertilisers shall only be used in a supplementary role to carbon based materials. Permission for use shall only be given when other fertility management practices have been optimised.
- The certification programme shall lay down restrictions for the use of inputs such as mineral potassium, magnesium fertilisers, trace elements, manures and fertilisers with a relatively high heavy metal content and/or other unwanted substances.

- Chilean nitrate and all synthetic nitrogenous fertilisers, including urea, are prohibited.

Pest, Disease and Weed Management

- Products used for pest, disease and weed management, prepared at the farm from local plants, animals and micro-organisms are allowed.
- Thermic weed control and physical methods for pest, disease and weed management are permitted.
- All equipments from conventional farming systems shall be properly cleaned and free from residues before being used on organically managed areas.
- The use of synthetic herbicides, fungicides, insecticides and other pesticides is prohibited.
- The use of synthetic growth regulators and synthetic dyes are prohibited.
- The use of genetically engineered organisms or products are prohibited.
- Accredited certification programmes shall ensure that measures are in place to prevent transmission of pests, parasites and infectious agents.

Contamination Control

- In case of reasonable suspicion of contamination, the certification programme shall make sure that an analysis of the relevant products to detect the possible sources of pollution (soil and water), shall take place to determine the level of contamination.
- For protected structure coverings, plastic mulches, fleeces, insect netting and silage rapping, only products based on polyethylene and polypropylene or other polycarbonates are allowed. These shall be removed from the soil after use and shall not be burnt on the farmland. The use of polychloride based products are prohibited.

Soil and Water Conservation

- Clearing of land through the means of burning organic matter shall be restricted to the minimum.
- The clearing of primary forest is prohibited.
- Relevant measures shall be taken to prevent erosion.
- Excessive exploitation and depletion of water resources shall not be allowed.
- The certification programme shall require appropriate stocking rates which do not lead to land degradation and pollution of ground and surface water.
- Relevant measures shall be taken to prevent salination of soil and water.

Collection of Non Cultivated Material of Plant Origin and Honey

- Wild harvested products shall only be certified organic if derived from a stable and sustainable growing environment. Harvesting or gathering the product shall not exceed the sustainable yield of the ecosystem, or threaten the existence of plant or animal species.

- Products can only be certified organic if derived from a clearly defined collecting area, which is not exposed to prohibited substances, and which is subject to inspection.
- The collection area shall be at an appropriate distance from conventional farming, pollution and contamination.

C. Food Processing and Handling

- Organic products shall be protected from co-mingling with non-organic products. All products shall be adequately identified through the whole process.
- The certification programme shall set standards to prevent and control pollutants and contaminants.
- Organic and non-organic products shall not be stored and transported together except when labelled or physically separated.
- Certification programme shall regulate the means and measures to be allowed or recommended for decontamination, cleaning or disinfection of all facilities where organic products are kept, handled, processed or stored.
- For pest management the following measures shall be used in order of priority.
- Preventive methods such as disruption, elimination of habitat and access to facilities.
- Mechanical, physical and biological methods.
- Approved Pesticidal substances and other substances may be used in traps.
- There shall never be direct or indirect contact between organic products and prohibited substances. (e.g. pesticides). In case of doubt, it shall be ensured that no residues are present in the organic product.
- Persistent or carcinogenic pesticides and disinfectants are not permitted. The certification programme shall set up rules to determine which protection agents and disinfectants may be used.
- In cases where an ingredient of organic agriculture origin is not available in sufficient quality or quantity, the certification programme may authorise use of non organic raw materials subject to periodic re-evaluation. Such non-organic raw material shall not be genetically engineered.
- The same ingredient within one product shall not be derived both from an organic and non-organic origin.
- Minerals (including trace elements), vitamins and similar isolated ingredients shall not be used.
- The certification programme may, grant exceptions where use is legally required or where severe dietary, or nutritional deficiency can be demonstrated.
- Preparations of micro-organisms and enzymes commonly used in food processing may be used, with the exception of genetically engineered micro-organisms and their products.

EMERGING ISSUES AND FUTURE DIRECTIONS IN ORGANIC CERTIFICATION

Organic certification has emerged into a highly specialized and critical functional area of the marketing of organic produce. There are many challenges and grey areas that are yet to be addressed adequately. The inter acceptability and comparability of certificates issued by different certification agencies is a key issue in certification. The high cost of certification, which makes it unattractive for small producers is another concern. The cost aspect of certification is highly relevant to Indian conditions as the average size of the holding are relatively low. The move towards a universally acceptable single standard for certification is the ultimate aim. But given the geo-political nature of trade and multilateral agreements, this may not be feasible in the near future. The demand for organic products has increased over the years. This has lead to increasing numbers and types of products and production systems and technologies are being labelled organic. This creates challenges for standard setters, certification bodies, regulators and accreditation bodies. The process of standard setting in organic products is an evolving area. This is evident from the many revisions and modifications which are being made in the standards already available.

The emerging trends like Internal control systems which allows an external certification body to delegate the periodical inspection of individual group members to an identified body or unit within the certified operator is aimed at overall cost reduction in the certification process. Such a system facilitates smallholder certification and helps in implementing and maintaining a high quality assurance system for organic standards in smallholder production. Group certification also enables smallholders to access organic markets and enables developing countries to commercialize their products at the international level.

The setting, control and proper administration and regulation of organic standards was key to the long term expansion of the market for organic produce in many developed countries. This holds important lessons for the developing countries to develop a vibrant market for organic product domestically and for developing capability in export of organic produce. The adherence to standards and a reliable certification process is sine qua non for developing and maintaining consumer confidence in organic products and thereby promoting the growth of the market for organic produce.

REFERENCES

Padel, S. and N. Lampkin 1994. Conversion to Organic Farming: An Overview. In The Economics of Organic Farming: An International Perspective. N. Lampkin and S. Padel (eds). Wallingford, *CAB International*: 295-313.

www.dacnet.nic.in/rbdcnagpur/organic_farming.htm

www.fao.org

www.ifoam.org

❑❑❑

Chapter **5**

Organic Farming : Status and Scope in Special Reference to Indian Agriculture

Archana Singh

Department of Botany, M.S.J.Government Post Graduate College, Bharatpur (Rajasthan)-321001

Organic farming is a form of agriculture which avoids or largely excludes the use of synthetic fertilizers and pesticides, plant growth regulators, and live stock feed additives. As far as possible, organic farmers rely on crop rotation, crop residues, animal manures and mechanical cultivation to maintain soil productivity and health to supply plant nutrients, and to control weeds, insects and other pests.

According to the International Organic Farming Organization (IFOAM): "The role of organic agriculture, whether in farming, processing, distribution, or consumption, is to sustain and enhance the health of ecosystems and organisms from the smallest in soil to human beings." Approximately 31 million hectares (75 million acres) worldwide are now grown organically.

Modern agriculture largely depends on the use of chemical fertilizers, pesticides, and herbicides. Such modern technologies have undoubtedly increased production and labour efficiency, which shows their deleterious effects on soil productivity and environmental pollution. Organic agriculture excludes the use of certain synthetic inputs, such as synthetic fertilizers, pesticides, herbicides and genetically modified organisms (GMOS). Methods of organic farming vary; however, organic approaches share common goals and practices. In addition to the exclusion of synthetic agrochemicals include the protection of soil like soil erosion, nutrient depletion etc. It also promotes the biodiversity and outdoor grazing for livestock and poultry.

The emerging global prospect for organic agriculture is the realization of long lasting damages caused by chemical fertilizers and pesticides both in developed and developing countries. World wide, about 130 countries produce certified organic products on commercial scale. The global market for organic food is worth $38 billion. The organic agriculture can contribute to socio-economic and ecologically sustainable development, especially in developing countries. The market for organic products is in a growing stage and offers producers and exporters to improve their incomes and living conditions. But what is the status of organic agriculture in developing countries. Organic agriculture is developing rapidly and is now practiced in more than 130 countries of the world. Globally 30.4 million ha of the agricultural land, is currently managed organically. India ranks 33rd in total area under organic cultivation. India has now become a leading supplier of Organic Herbs-2008. It was around 78 million US$ registering a 200% growth over the previous year. Though organic farming is an inevitable phenomenon, it alone cannot feed the billion population of India (Mahapatra *et al*, 2009).

By the starting of green revolution in India it was mainly realized that with the introduction of high yielding varieties of various crops and by following intensive cultivation practices with the use of fertilizers, pesticides and other inputs the production could be increased. The fertilizers consumption in the country accordingly has increased from 66,000 in 1951-52 to about 12.16 million tones in 1992-93. At present the consumption of fertilizer in India is 26.3 million tones. Same is the situation with pesticides. All these agrochemicals are produced from fossil fuels.

Reports of deleterious effects by degradation of lands and increasing chemical pollution by agrochemicals were being observed by Government of India. In order to balance such situation organic agriculture is gaining gradual popularity across the world. The demand for domestic consumption is increasing in many countries including India. Australia has the maximum area under organic agriculture, followed by Europe and the American continent. The growth rate of organic market is highest in Japan, USA, Australia and the European countries.

India has the largest arable land in the world and second largest producer of fruits and vegetables. Instead of all the favorable factors we are still struggling in regard of agriculture in the world. By default farmers adopt organic farming in many parts of India especially due to high cost of pesticides and fertilizers which is the another reason for conversion to organic agriculture. Moreover, 65% cultivated area in India is rainfed where the use of chemical fertilizers and pesticides is negligible. The abundant availability of poor labour also suits to India to follow the path of labour intensive organic agriculture. While organic cultivation is taken up in 76,326 hectares.

DEFINITION AND OBJECTIVES OF ORGANIC FARMING

The Organic farming in real sense is the management approach to improve the health of underlying the productivity of soil (Palaniappan and Annadurai, 1999).

According to Lampkin (1999). The term "Organic" is best thought of as referring not to the type of inputs used. But to the concept of the farm as an organisms in which all the components, the soil minerals organic matter, microorganisms, insects, plants, animals, humans interact to create coherent. Self regulating and stable whole reliance on external inputs whether chemical or organic is reduced as far as possible organic farming is (w)holistic production system.

Thus the Organic farming is directly related to soil fertility as a key to successful production.

HISTORICAL PERSPECTIVE OF ORGANIC FARMING

- Ancient Period: Oldest practice 10000 years old, dating back to Neolithic age.
- Ramayana: All dead things- rotting corpse or stinking garbage returned to earth are transformed into wholesome things that nourish life.
- Mahabharata: 5500BC mention of Kamadhenu, the celestial cow and its role on human life and soil fertility.
- Kautiliya Arthashastra: 300BC mentioned several manures like oil cake, excreta of animals.
- Brihad-Samhita: By Varamihir described how to choose manures for different crops and the methods of manuring.
- Rig-Veda: (2500-1500BC) Mention of Organic Manure in Rig-Veda 1, 161, 10.
- Atharvaveda: (1000BC), Green Manure II, 8.3.
- Sukra (IV, V, 94, 107-112) it is stated that to cause healthy growth the plant should be nourished by dungs of goat, sheep, cow, water as well as meat.
- Vrikshaayurveda: A reference of Manure, No, 324 B, six, 107-164.
- Holy Quran (590AD): At least one third of what you take out from soils must be returned to it implying recycling or post harvest residue.
- 1924: Rudolf Steiner's spiritual foundations in Germany for the renewal of agriculture, led to the popularization of biodynamic agriculture.
- 1940: The first term Organic Farming used by Lord Northbourne. The term is derived from his concept of "The Farm as Organism". In his book, look to the land, he described a holistic, ecologically balanced approach to farming.
- 1940: The British Botanist, Sir Albert Howard studied traditional farming practices in Bengal, India. He adopted Northbourne's terminology in his book "The Soil and Health: A study of Organic Agriculture" in 1947.
- 1940: In the US J. I. Rodale popularized Organic gardening among consumers.
- 1940: The Japanese farmer and writer Masanobu Fukuoba invented a no-till system for small scale grain production that he called natural farming.
- 1972: The International Federation of Organic Agriculture movements (IFAOM), was founded in Versailles, France. IFAOM was dedicated to the diffusion of information on the principles and practices of Organic Agriculture across National and linguistic boundaries.
- 1980: Various farming and consumer groups worldwide began pressing the government regulation of organic production.
- 1990: Legislation and certification standards were being enacted begins.

MAIN OBJECTIVES OF ORGANIC FARMING

The main objectives of organic farming are as follows:

1. To avoid pollution occur from hazardous chemical fertilizers and pesticides.
2. Improve the long term soil fertility by using organic manures.
3. Improve the best nutritional qualities of foodstuffs.
4. To reduce the use of fossil energy in agricultural practice to a minimum.

ORGANIC FARMING PRINCIPLES AND PROTOCOLS

Organic Farming emphasizes using only organic fertilizers for fertility maintenance and banning nearly all chemical pest control methods. However, Gypsum, an inorganic calcium salt and rock phosphate is allowed as fertilizer, and a number of natural sprays are also allowed. Organic farmers can use Anti-biotic derived from fermentations like streptomycin. To encourage the organic agriculture, Government has passed laws to create a demand for organic foods. Cities have created organic markets where organic producers can sell their organic items.

Why we do for Organic Agriculture

Inadequate knowledge and short sightedness in using the chemical fertilizers and pesticides are silently and extremely harmful to our health. They easily enter into the food supply and seriously harm the livestock. They penetrate in soil, water sources and disturb the ecological balance. So we can easily access and evaluate the situation and usage of chemical fertilizers and pesticides. And can be get rid of by promoting the organic agriculture by using various healthy and sustainable farming practices. India's organic agriculture is about gaining wisdom of our agriculture heritage. It vitalizes and enriches the soil in every season by increasing the long term yield of crop and nutrient value. Organic agriculture is the best social method for farmers, their families, livestock, farmlands and communities.

Consciousness about health and conservation of environment has brought a major shift towards organic agriculture. Globally people looking forward to organic food that is considered safe and hazard free. Thus the use of synthetic fertilizers and pesticides is avoided. The global market for organic food is expected to touch US$ 29-31 billion by 2005. Day by day the demand for organic food is increasing both in developed and developing countries. Approximately 130 countries are engaged in producing certified organic products commercially (Kortbech-Oksesen, 2000).

CHARACTERISTICS OF ORGANIC FARMING

- Protection of fertility of soils by maintaining organic matter levels for long terms, encourage oil biological activity and mechanical intervention.
- Use of insoluble nutrients made available by the action of soil micro-organisms.
- Use of legumes for nitrogen self sufficiency, biological nitrogen fixation and recycling of organic materials including livestock manures and crop residues.

NUTRIENT MANAGEMENT IN ORGANIC AGRICULTURE

Organic agriculture is the form of agriculture includes crop rotation, application of green manure, compost, biological pest control to maintain soil productivity and restrict the use of synthetic fertilizers and pesticides. Crop rotation and green manure help to provide nitrogen for leguminous crop (Fabaceae family) which fix nitrogen from the atmosphere through symbiosis of bacteria Rhizobium intercropping of the plants. It is helpful to control pests and insects and also improve the soil nutrients (Watson *et al* 2002).

In organic agriculture we use animal manure compost, processed fertilizers, rock phosphate and green and naturally occurring form of potash which provides potassium. Some synthetic compounds such as Iron Sulphate, Aluminum Sulphate and Soluble Boron products are allowed in organic agriculture in the U.S (Gillman, 2008).

Infact, the organic agriculture rely on the management of the soil. It enhances the chemical, biological and physical properties of the soil. The plant nutrition depends on "Biologically-derived nutrients instead of using readily soluble forms of nutrients. The increasing amount of nutrition in soil plays key role in suppressing weeds, pests and diseases (IFOAM, 1998).

Animal dung, crop residues, green manures, bio-fertilizers and food processing waste products are some of the common source of Organic nutrition. The several compost production technologies like Vermin-composting, Phospho-composting, Nitrogen rich Phospho-composting etc., improve the soil quality by enriching the nutrients. These manures have the capacity to fulfill nutrient demands of crops and promote the activity of beneficial macro- micro flora of the soil (Singh, 2003).

The basic requirement in organic farming is to increase input use efficiency at each step of the farm operations. This is achieved partly through reducing losses and adoption of new technologies to enrich the nutrient content in manure (AICRP).

The combination of vermin-compost with farm-yard manure enriches the nutrients three to four times in Organic Agriculture. Vermin-composting, use of biomass raised on the bunds as source of nutrients and bio-fertilizers are best approaches that were tried in organic farms in the country. Adoption of soil and water conservation measures, a key component of organic agriculture. Mulching residue management, green leaf manuring, cover cropping are other strategies that conserve moisture and improve nutrient efficiency in organic production methods. The use of farmyard manure (FYM) or other organic nutrient sources during aberrant rainfall years have additional advantages of protecting crop from drought besides the nutritional benefits.

Under USDA organic standards, manure must be subjected to proper thermophilic composting and allowed to reach a sterilizing temperature. If raw animal manure is used 120 days must pass before the crop is harvested if the final product comes into direct contact with the soil. For products which do not come into direct contact with soil, 90 days must pass prior to harvest.

GROWTH, ECONOMICS AND MARKETING OF ORGANIC FARMING

As of 2001, the estimated total market value of certified organic products was estimated to be \$20 billion. By 2002 this was \$23 billion and by 2007 more than \$46 billion according to organic monitor (Willer and Kilcher, 2009).

In recent years both Europe (2007:7.8 million hectares/ European Union: 7.2 million hectares) and north America (2007:2.2 million hectares) have an experience of strong growth in organic farmland. However, this growth has occurred under different conditions. While the European Union has shifted agricultural subsidies to organic farmers in recognition of its environmental benefits, the United States has taken a free market approach (Dimitri and Oberholtzer, 2006).

As a result, as of 2008 4.1 percent of European unions farmland was organically managed compared to just 0.6 percent of United States farmland (Willer and Kilcher, 2009). The growth of organic farmland area in the EU was 21% from 2005 to 2008 reaching a total of 7.8 million hectares.

IFOAM's most recent edition of "The world of Organic Agriculture: Statistics and Emerging Trends 2009 lists the countries which had the most hectares in 2007. The country with the most organic land is Australia (12million hectares), followed by Argentina, Brazil and the US.

The economics of organic farming is related to the effects in terms of human society. Organic farming is labour and knowledge intensive whereas conventional farming is capital intensive, requiring more energy and manufactured inputs (Strochlic and Sierra, 2007).

The replacement of external inputs by farm derived resources normally leads to reduction in variable input costs under organic management. Expenditure on fertilizers and pesticides is lower than in conventional systems in almost all the cases. (Offer Mann and Neiberg, 1999; Padel and Lampkin, 1994).

Studies have shown that the common organic agricultural combination of lower input costs and favourable price premiums can offset reduce yields and make organic farms equally and often more profitable than conventional farms (Peterson *et al* 1999; Reganold *et al* 2001; Hanson *et al*, 1997).

According to the marketing requirement the certified as organic is often a very useful distinction that helps to differentiate an organic farmer's products from the conventional competition. Marketing is often a constraint for any kind of farmers and it can be especially difficult for organic farmers. As organic products increasingly find their way into the most common market channels, including supermarkets, the discovery process becomes easier while the demand become more difficult. In Uttaranchal, the attempt of a loose conglomeration of farmers to deal directly with an urban retailer ended poorly because neither they nor the government agents facilitating the transaction were familiar with the business requirements of this trade. A number of cases learned to use traders to facilitate such transactions since small farmers are typically constrained in three distinct areas of marketing. First, farmers should assess their specific market orientation by honestly evaluating what they have to offer. Second, farmers must learn the requirements needed to access their targeted organic markets. By mapping out market channels, they can better understand purchasing patterns and behaviour so as to ascertain the current and future market potential and its attractiveness. Third, farmers must recognize that these processes require dedicated attention and some training.

According to demand in the domestic markets, the local demand for organic products has grown along with consumer's incomes and their increased concern for food safety. India's domestic market is small and mostly informal with only a few shops dedicated to organic products. Much of the organic produce reaches consumers without being subject to organic identification or specific labeling. Therefore there are no credible or complete estimates for the size of the domestic market. One survey notes that more organic products in India are sold through the supermarket channels (31%) and to the processing industry (30%) than through any other (Garibay and Jyot, 2003). Most of the organic food retailers in India are located in urban areas where food availability and sales have outpaced the growth of rural food supplies. These more affluent consumers have the disposable income to afford the higher price of organic products but sales volumes are still quite modest. This is in part due to:

1. Modest availability and selection in stores exacerbated by limited prominence
2. Inconsistent supply from farmers
3. Sometimes exorbitant prices
4. Poor consumer understanding of organics

According to the Indian competence centre for organic agriculture, the global market for organically produced foods is $26 billion and is estimated to increase to $102 billion by 2020. As part of 10th five year plan (2002-2007), the government earmarked Rs. 100 crore (Rs. 1 billion) for the promotion of sustainable agriculture in the country, but the main components of this initiative have beneficial exports, from the establishing of national organic standards under NPOP (National programmed for organic production), putting in place a system of certification for products, and establishing APEDA (agricultural and processed food export development authority) as the nodal agency to promote exports opportunities. When organic farmers and traders are operating in an anonymous market, certification is developed to show and guarantee to consumers that a product has been produced in consistency with organic standards. Under current government policy, it takes approximately three years for a farm in conversion to be certified as organic, and costs are hefty for the small scale farmer. While internationally, our exports cannot hold ground without certification, uncertified brands in India have had some success.

In developed countries, especially in Europe and the US, every supermarket has a green line where a complete range of certified organic products is available. India is just beginning to manage a toehold in some supermarkets.

Fabindia is hoping to change that. The Rs. 100 Crore (Rs. 1 billion) apparel and home furnishings retailer ventured into organic foods earlier this year when it test-marketed its organic range in New Delhi.

India is bestowed with lot of potential to produce all varieties of organic products due to its various agro climatic regions. In several parts of the country, the inherited tradition of organic farming is an added advantage. This holds promise for the organic producers to tap the market which is growing steadily in the domestic market related to the export market.

Currently, India ranks 3 rd in terms of total land under organic cultivation and 88th position for agriculture land under organic crops to total farming area. The cultivated land under certification is around 2.8 million hectare (2007-2008). This includes 1 million hac under cultivation and the rest is under forest area (wild collection). The government of India has implemented the national programme for organic production (NPOP). The national programmed involves the accreditation programmed for certification bodies; norms for organic production, promotion for organic farming etc. the NPOP standards for production and accreditation system have been recognized by European commission and Switzerland as equivalent to that of US. With these recognitions, Indian organic products duly certified by the accredited certification bodies of India are accepted by the importing countries.

India produces around 3,96,997 MT of certified organic products which includes all varieties of food products namely basmati rice, pulses, honey, tea, spices, coffee, oil seeds, fruits, processed food, cereals, herbal medicines and there value added products. The production is not limited to the edible sector but also produces organic cotton fiber, garments, cosmetics, functional food products, body care products etc.

India exported 86 items in 2007-08 with the total volume of 37533 MT. The export realization was around 100.4 million US$ registering a 30% growth over the previous year. Organic products are mainly exported to EU, US, Australia, Canada, Japan, Switzerland, South Africa and Middle East. Cotton leads among the products exported 16,503 MT. (APEDA, 2008)

USE OF FERTILIZERS IN ORGANIC AGRICULTURE

The fertilizers needed for cell growth, protein synthesis, root development and good biomass etc. (Bhattacharya, 2005). First fertilizer industry of Single Super phosphate (SSP) comes into the light in 1842 at Roth Amsted Experimental Station, London. In India it originates in 1906 in Thailand. The world food grain production is to the extent of 2100 million tones roughly which is generally based on farming where adequate use of fertilizers and pesticides are in practice.

Worldwide, plant nutrient consumption increased from 27.4 million tones N+P2O5+K20 in 1960 to 149.8 million tones in 2004-2005. Fertilizer consumption is projected to amount to 163 million tones N+P2O5 + K2O in 2008/2009 and 208 million tones in 2020. Phosphate fertilizer consumption increased from 6.2 to 37 million tones P2O5 from 1960 to 2005 and is projected to amount to about 39.5 million tones P2O5 in 2015. Trends in fertilizer consumption show that from 1960 to the early nineteen's developed countries accounted for more than half of world fertilizer use. In 1990, the fertilization consumption was about 80 million tones nutrients. The projection is by the year 2020, fertilizer consumption in developing countries will be at least 40% higher than the amount of fertilizer consumption in developed countries.

The steadily increase in fertilizer use in developing countries are much more impressive in Asia, which emerged as the region that uses about 50% of world fertilizer consumption (Figure-1). Over the past decades, countries such as china, India, Pakistan and Vietnam witnessed tremendous increases in their consumption of fertilizer.

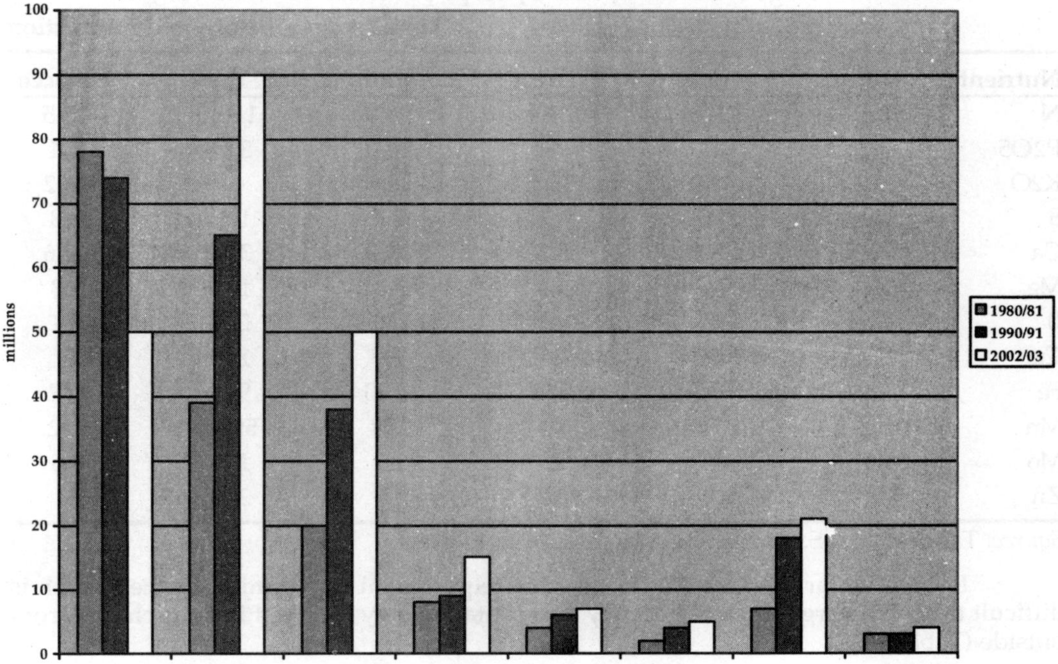

Fig. 1 : Total fertilizer consumption trends in main regions of the world

Here:

DEPD COUN=Developed countries; DEVG COUN=Developing countires; E&SE, CH, JP= East & South East Asia, China, Japan; LA=Latin America; NE=Near East; NA=North Africa; SA=South Asia; SS&SA=Sub Sahara & South Africa

Source: FAO, 2006

These countries show also drastic increases in agricultural production; with some of them turning from huge food deficits and famines to the situation of exporters of grains (the case of India is well known). Trends in fertilizer use and production in India were as follows (Table-1).

Table 1 : Fertilizer use and food grains (Mt) productions in India, 2001-2006

Year	Total Nutrient Consumption (N+P2O5+K2O)	Total Food grains Production
2001-2002	17.4	212.9
2002-2003	16.1	174.8
2003-2004	16.8	213.2
2004-2005	18.4	198.4
2005-2006	20.3	208.3

There is scientific evidence that plant uptake inorganic nutrient and crop has its demand (Table-2)

Table 2 : Average Nutrient uptake by crops per tones of economic yield

		Uptake per tone of yield production				
Nutrient	**Unit**	**Paddy**	**Wheat**	**Groundnut**	**Tea**	**Tobacco**
N	Kg	20.0	25.0	58.0	110.0	16.5
P2O5	Kg	11.0	9.0	20.0	37.8	4.4
K2O	Kg	30.0	33.0	30.0	4.0	26.2
S	Kg	3.0	4.7	5.7	10.1	2.8
Ca	Kg	7.0	5.3	28.0	31.0	22.6
Mg	Kg	3.0	4.7	7.3	10.7	4.8
B	Kg	15	48	133	200	96
Cu	Kg	18	24	15	240	11
Fe	Kg	153	624	1500	900	692
Mn	Kg	675	70	118	5900	132
Mo	Kg	2	2	4	Na	0.6
Zn	Kg	40	56	28	240	21

Source: Tandon, H.L.S. Fertilizer Guide.21 (2002).

In organic farming implies large scale requirement of organic fertilizer which is difficult to obtain. Organic fertilizer may also contain heavy metals if it is purchased from outside (Table-3).

Table 3 : Total concentration of selected heavy metals in manures (ppm as dry wt. basis)

Source	Arsenic	Cadmium	Chromium	Lead	Nickel	Copper
Cow manure	-	08.0	58	16	29	62
Poultry manure	0.35-110.5	-	0.6-19.6	-	-	3.5-13.5

Source: Chhonkar, P. K. J. Soil. Sci. Soc. India, Vol.51, No. 4, 365-377 (2003).

The use of chemical fertilizer is not good for health because they contain heavy metals which cause several diseases due to excessive nitrates and create environmental pollution (Table-4).

Table 4 : Concentration of selected heavy metals (ppm on dry wt. basis)

Sources	Arsenic	Cadmium	Lead	Nickel
Urea	< 0.04	< 0.2	< 0.4	< 0.2
DAP	9.9-16.2	4.6-35.5	2.1-3.7	7.4-222
MOP	0.4	< 0.2	< 0.4-10	< 0.2
TSP	10.3	15.0	11	17

Source: Chhonkar, P. K. J. Soil. Sci. Soc. India, Vol.51, No.4, 365-377 (2003).

Table 5 : Adverse effects of nitrogenous fertilizers on human health and environment

Effect	Causative Agents
Human Health	Excess NO3 and NO2
Methmoglbinemea Cancer	Nitrosamine illness from NO2, Secondary Amines, Peroxyacyl Nitrate
Environmental Health	Excess NO3 in feed and water
Eutrophication	Inorganic and organic N in surface water
Materials and Ecosystem damage	HNO3, Aerosols in rainfall

Source: Bhattacharya, P. 2004. Organic Food Production in India Status, Strategy and Scope. Agribios (India) Jodhpur 01-182 (2004).

Gradually by the use of chemical fertilizer the compound annual growth rate of important crops in India is going to decline (Yadav, 2003). Many studies all over the world have done a simple regression analysis between food grain production and fertilizer consumption during 1961 to 2006 and showed that the partial factor productivity of fertilizers has been continuously declining. The data indicate a reduction in crop response to fertilizer application, especially when balanced fertilization is not practiced. Furthermore, the soils are also getting continuously depleted of secondary plant nutrients and micronutrients. If this nutrient drains continuously, the sustained high productivity and sustainability of agriculture will be an impossible task. Thus, inadequate and imbalanced fertilization is a major causative factor for low efficiency of fertilizer use in crop production. Nutrient consumption ratios of some regions and the world are present in the Table 6.

Table 6 : Nutrient consumption ratio (2002)

Countries	N	P2O5	K2O
World	1.0	0.4	0.2
Developed countries	1.0	0.4	0.3
Developing countries	1.0	0.4	0.2

contd. ...

... contd.

Countries	N	P2O5	K2O
Latin America & Caribbean	1.0	0.7	0.6
Africa	1.0	0.3	0.2
East & south east Asia	1.0	0.3	0.3
South Asia	1.0	0.3	0.1
Europe	1.0	0.2	0.3

In the other regions, from Africa through Europe and Asia, P and K applications are much unbalanced with regard to N. data from countries of Asia, where IMPHOS is being conducting several promotional activities towards balanced use of fertilizers, are presented below. In India, data from 1980-1981 to 2005-2006 are given in table-7.

Table 7 : Ratio of use of NPK, India

Year	N	P2O5	K2O
1980-1981	1.0	0.3	0.1
1985-1986	1.0	0.4	0.1
1990-1991	1.0	0.4	0.2
1992-1993	1.0	0.3	0.1
1993-1994	1.0	0.3	0.1
2000-2001	1.0	0.4	0.1
2005-2006	1.0	0.4	0.2

Among many consequences that resulted from unbalanced use of fertilizers is as follows:

- The decrease in fertilizer use efficiency, in particular, for Nitrogen in china, results from a study covering 5.7 million hectares shows that on an average fertilizer use efficiency decreased by 8.2%;
- Yield reduction that, in the above example ranged from 0.4 to 1 t/ha;
- Reduction of farmer's income;
- The mining, in some situation (India) of P and K on a very large scale;
- Increasing N losses to the environment by leaching and volatilization;

But on the other hand the use of organic sources plant also uses nutrients by activity of microorganisms which available in soil. The organic nitrogen converted to nitrates by *Nitrosamines* and *Nitrobacteria, Bacillus* and *Pseudomonas* etc. Organic crops contain more nutrients like vitamins minerals, enzyme, trace elements and anti-oxidants. Initially at the starting the organic farming produce less yields but later it increases rapidly. Regular addition of organic fertilizer improves the soil quality (Table-5).

Table 8 : Mean value of aggregated soils data from 16 pairs of farms each with organic (Bio) and conventional (Con) farming

Soil Property	All "Bio" farm	All "Con" farms
Bulk density (mgm-3)	1.07	1.15*
Penetration Resistance (0-20cm) (Mpa)	2.84	3.18*
Carbon %	4.84*	4.27
Respiration (ul O2 l-1g-1)	73.7*	55.4
Mineralizable-N (mg kg-1)	140.0*	105.9
Ratio of Mineralisable N TO C (mg g-1)	2.99*	2.59
CEC(C mol kg-1)	21.5*	19.6

* = Significantly different at p< 0.01

** Cation Exchange Capacity in Centimoles of Cation Charge (=) per kilogram of soil

Source: Regan old J. P, Palmer A. S. 1993. Lockhart, I. C; and Macgregor, A.N. Science, 260: 344-349.

Due to scanty awareness of organic benefits the farmers like to go for conventional farming for their profit. So organically grown products are expensive due to less percentage of organic farming in India. It is needed to promote the organic agriculture so the goods can purchase by every man of the country. In the present scenario every man is aware and has high demand of organic food.

Phosphate Fertilizers and Food Quality

The quality of vegetable products depends on many factors, among others, optimal supplies of all substances required for growth. Minerals occupy a central position for obtaining high quality products.

Properly applied fertilization improves food quality through a higher quality of vegetable products (thus indirectly of animal products, as well) so it contributes not only to nourish men and animals but also to provide them with healthier living conditions.

Many important tasks of phosphate in metabolism are the reason that P supplies play a central role in food quality. Phosphate supplies cause the following changes:

Use of Potassium Bicarbonate as a Fungicide in Organic Farming

Potassium bicarbonate can be used as a contact fungicide in a variety of crops, e.g. grapevine, pome and stone fruit, berries and soft fruit, vegetables and cereals. In Europe, efficacy trials are still underway at present. It is applied with standard array equipment at a rate of 2 to 8 kg/ha, with a maximum of 8 applications per year. It mainly inhibits fungus mycelium development. Its mode of action is linked with osmotic pressure, pH and specific bicarbonate/ carbonate ion effects. In some crops, potassium bicarbonate has the potential to replace copper or sulfur fungicides. Potassium bicarbonate is made industrially from carbon dioxide (CO2) gas and potassium hydroxide (KOH).

Potassium bicarbonate is naturally present in humans, animals, plants and virtually all living organisms. Under environmental conditions, potassium bicarbonate dissociates completely to potassium and bicarbonate ions and it is impossible to differentiate between ions naturally present and those of external origin. Bicarbonate is present in soil pore waters as a result of carbon dioxide liberated from the respiration of soil organism. Potassium is an essential plant and microbial nutrient that has a natural cycle in soil of uptake and utilization by plants and microbes, followed by release resulting from the decomposition of rotting organisms. Potassium input resulting from use as a fungicide is considerably smaller than the crops potassium needs. Further, potassium bicarbonate has an extremely low toxicity to mammals and is not hazardous to the environment.

In relation to the human health because of its extremely low toxicity, potassium bicarbonate presents no health risks to operators and by standards, and its residues present no risks to consumers. It is a recognized food additive (E 501).

Potassium bicarbonate has been used for decades in medicine and everyday products (as a food ingredient and as a leaving agent in baking). Thus, it is perceived as a safe substance by consumers. In the farming community, the partial replacement of copper fungicides by potassium bicarbonate will be perceived favourbly.

Baking powder (sodium bicarbonate) has been used by organic farmers for decades, and is still used as plant strengthener in Germany, potassium bicarbonate is allowed for use in organic farming in the USA, and also by the IFOAM standards.

Thus potassium bicarbonate occurs in nature, is an effective fungicide, and is safe for humans and the environment. As a fungicide, it has the potential to replace copper and sulphur in some crops. However, for commercial use it has to be synthesized. Thus the potassium bicarbonate will be recommended as a fungicide in organic farming.

STANDARDS, ACCREDITATION AND CERTIFICATION OF ORGANIC FARMING

Standards are the guiding principles for organic farming and processing practices. At present, there are no regulations on organic products applicable world wide. Recently, attempts have been made by FAO/WHO Codex Alimentations Commission to harmonies the laws and to establish an international procedure to develop the equivalence of organic standards. Two international agencies IFOAM and Codex Alimentarious commission are the prime players in framing standards. In the EC, regulation EEC 2092/91 determines the minimum requirements for organic farming in all the member states. It contains standards for production, processing, impacts, inspection and certification, marketing and labeling of organic products. In India, the national programme for organic production (NPOP) provides an institutional framework for implementing standards for organic production system. This includes certification of organic farms, products and processes as per national standards of organic products (NSOP) (K. Usha. 2006. Technical Digest)

Globally there are more than 60 standards which include IFOAM basic standards, CODEX Alimentarious Commission guidelines, EU regulation 2029/91. NOP of USA etc (Table-9).

Table 9 : Some facts on international standards

1.	IFOAM	- Established in 1972
		- Headquarter in Germany.
		- Umbrella Organization for organic agriculture association.

contd. ...

... contd.

		- Developed International Basic Standards of Organic Agriculture. - Established IFOAM Accreditation Program (1992) to accredit certifying bodies. - Set up International Organic Accreditation Service (IOAS) in July 2001.
2.	CODEX	- Codex Alimentarious Commission – A joint FAO/ WHO. - Inter Government Body-Established in 1962. - Produced a set of guidelines for organic production.
3.	EU Regulation	- Laid out a basic regulations for European Union's Organic Standards in Council Regulation No. 2092/ 91 (June, 1991). - Regulations give guidelines for the production of Organic crops in the European Community.
4.	Demeter	- Demeter International is a world wide net work of 19 International Certification bodies in Africa, Australia, and Europe. - Developed guidelines for Biodynamic preparation.
5.	JAS	- A set of guidelines "Japan Agricultural Standards" for organic production.

Source: IFOAM- The World of Organic Agriculture Statistics and Emerging Trends. (Helga Willer and Minou Yussefi eds). 1-167 (2004).

As of 2007 over 60 countries have regulations on organic farming (IFOAM 2007:11). In 2005 IFOAM created the principles of organic agriculture, International guidelines for certification criteria. Typically the agencies do not certify individual farms, but rather accredit certification groups. Materials used in organic production and foods are tested independently by the organic materials review institute.

Some of the Important Organic Production Requirements as per National Standards for Organic Production developed by APEDA are reproduced Below

1. Genetically engineered cultivars or plant materials are not permitted in organic production.
2. The seed for raising a crop should either be organically produced or if organic seed is not available, conventional seed without any chemical treatment may be used.
3. Whole farm including the livestock should be converted to organic in a step by step manner.
4. If the whole farm is not converted, the certification programme shall ensure that the organic and conventional parts of the farm are separate and inspectable.
5. Before products from a farm/project can be certified as organic, inspection shall be carried out during the conversion period.

6. To ensure a clear separation between organic and conventional production, the certification programme (agency) shall inspect, where appropriate, the whole production system.
7. Plant products produced can be certified organic when the national standards requirements have been met with during the conversion period of at least two years before sowing for annual crops or in the case of perennial crops other than grasslands, at least three years before the first harvest of products.
8. Biodegradable material of microbial plant or animal origin shall form the basis of the fertilization programmes.
9. Manures containing human excreta (faeces and urine) can not be used on vegetation for human consumption.
10. Mineral fertilizers shall only be used in a supplementary role to carbon based materials. Permission for use shall only be given when other fertility management practices have been optimized.
11. Chilean nitrate and all synthetic nitrogenous fertilizers, including urea, are prohibited.
12. Mineral fertilizers shall be applied in their natural composition and shall not be rendered more soluble by chemical treatment.
13. Products used for pest, disease and weed management, prepared at the farm from local plants, animals and micro-organisms, are allowed.
14. The use of synthetic herbicides, fungicides, insecticides and other pesticides is prohibited.
15. In case of reasonable suspicion of contamination the certification programme shall make sure that an analysis of the relevant products and possible sources of pollution (soil and water) shall take place to determine the level of contamination.
16. For protected structure coverings, plastic mulches, fleeces, insect netting and silage rapping, products based only on polyethylene and polypropylene or other polycarbonates are allowed. These shall be removed from the soil after use and shall not be burnt on the farmland. The use of polychloride based products such as PVC film is prohibited.

The detailed description of various regulations governing organic production can be obtained from APEDA or downloaded from their website www.apeda.com. These stringent production requirements have been laid down to help produce the food commodities suitable for export. According to recent estimates, organic food market is worth around 15 billion sterling pounds worldwide and growing at a fast pace. Therefore, development of production practices in accordance with regional conditions and farmer requirement is need of the hour. India can utilize her economical manpower to produce cost competitive and high quality products for export and for domestic market. Organic production is one area of agriculture which can convert India's "green revolution" into "evergreen revolution." However, it will require highest level of commitment from every individual and institution engaged directly or indirectly in promotion of organic farming. Who knows we might be looking at another billion dollar industry in the making similar to information technology.

Accreditation

It is the procedure by which an authoritative body evaluates and gives a formal recognition that a certification programme is in accordance with the standards of the authoritative body. The government of India has initiated the process of setting up indigenous certification programmes to solve the problems arising out of non existence of domestic certifying agencies and the high fee being charged by certifying agencies of foreign origin. As per the national accreditation policy under the national programme for organic production (NPOP) being implemented by the Government of India through the Ministry of Commerce, all the certifying agencies operating in the country have to obtain accreditation from anyone of the four accrediting agencies appointed by the government of India, viz, Spices Board, Coffee Board, Tea Board and APEDA for spices, coffee, tea and other agricultural crops. There is IFOAM accreditation programme, launched in 1992, by initiating International Organic Accreditation service (IOAS).

Certification

There are 364 certification bodies across the world (Table-10), 290 of them are located in European Union, USA, Japan, Canada and Brazil.

Table 10 : Certification bodies and their approval per region

Region	Total	IFOAM	Japan	ISO 65	EU	USA
Africa	7			1		
Asia	83	4	65	1	1	2
Europe	130	10	9	45	100	28
Latin America	33	4		10	5	8
North America	101	4	1	14		64
Oceania	10	4	6	3	6	4
Sum	364	26	81	74	112	106

Source: IFOAM

Certification of organic products is an essential prerequisite for marketing of products and to attract premium in the International Markets. Certification is one way of ensuring that the products claimed to be organic are actually produced according to organic farming principles. It is the way of protecting consumers, producers and traders against the use of misleading or deceptive labels. It is also a marketing instrument enabling producers to access markets for organic products and obtains premium prices. Certification tracks and controls the flow of products from primary production at farm level to each stage of manufacturing right to the final product that reaches the consumers. Broadly the certification process consists of:

a. Inspection to verify that production and handling are carried out in accordance with standards against which certification is to be done.

b. Certification to confirm that production and handling confirms the standards. The documents to be maintained includes field map, field history sheet, activity register, input record, output record, harvest record, storage record, sales record, equipment cleaning records, and labeling records. It may be the procedure to obtain certification for export by the certification bodies of the importing countries. Logos of these bodies are well known and trusted by consumers in their countries, thus giving the product a better visibility and commercial advantage. The major drawback is that this type of certification can be very expensive. In order to reduce costs to producers and exporters, International Certification bodies have started establishing branches in producing centres utilizing the services of local organic inspectors. International certification bodies such as ecocert have established local branches in producing countries. Organic certification can also be carried out by local certification bodies accredited by the authorities of the impoting countries.

THE CERTIFICATION AGENCIES OPERATING IN INDIA INCLUDES

1. Bioinspecta, Switzerland's branch: Bioinspecta, c/o Indocert, Aluva, Kerala.
2. Ecocert International (Germany), Ecocert SA Branch, Aurangabad Maharashtra.
3. Indian Organic Certification Agency (INDOCERT), Aluva, Kerala.
4. International Resources for Fairer Trade, Sona Udyog, Andheri (E) Mumbai.
5. IMO Control Private Limited, Banglore.
6. LACON GMBH Germany's branch c/o Renewable Energy Centre, Alwaye, Kerala.
7. Naturland-Association for Organic Agriculture, Germany's office India, Gurgaon, Haryana.
8. SGS India pvt. Ltd. Gurgon, Haryana.
9. Skal International (Netherlands Skal Inspection and Certification Agency, Banglore.

There are glaring examples like the village called Karunapuram in Idukki district of Kerala that has attained the status of cent percent Organic farming village, with all the 4,000 farmers in the village, committed to use only organic inputs for their farms. The Darjeeling district in west Bengal for Organic tea, Kodagu district in Karnataka for Organic coffee and Idduki district in Kerala for Organic spices. Uttaranchal, Sikkim, Nagaland and Meghalaya have declared themselves as organic farming states. At national level, Government of India has created an enabling environment for promoting organic farming and the supporting system set up. India has made a dent in the export of 31 Organic items which include farm produces like tea, coffee, spices, fruits, vegetables, cotton, rice, oilseeds, pulses, sugarcane and basmati rice. Currently, India accounts for one- third of the global Organic tea production. However, our share in the \$150 million global organic coffee market which is less than 1%. In the \$3.2 million organic spices market, India stake is just 3%. In view of the traditional advantages in these commodities, we hold the promise to substantially enhance the stake in the global trade of organic produce. This is requirement of the future that the rural areas must link with the export market.

CHALLENGES TO INDIAN AGRICULTURE

Thus the growing demand for food grains, vegetables, fruits, milk, poultry and meat as well as cash crops will present newer challenges to organic agriculture (Table-8). Let us not forget that our existing food security has been mainly brought about by the increase in irrigated agriculture and the introduction of high yielding varieties of crops. Current stability in production is through wheat, largely a winter crop. However, the rain fed areas, which account for 70% of the cultivated area of the country, have not benefited from the modern developments in organic agriculture. Of this 70% about 30% area is under dry land agriculture where annual rainfall is only up to 400 mm.

The problem in areas with rain fed agriculture need to be understood. Lesser the rain in an area, greater is the trouble for the farmers and villagers. During many seasons, it was rare to find even blades of grass. Now in some areas in arid Rajasthan, one is struck by the change brought about by irrigation waters of the Indira Gandhi Canal. The change in the quality of the people's lives is something that gives immense satisfaction and one visualizes India with many such canals- big and small-connecting different river systems and water bodies, and well managed water harvest and watersheds, thus benefiting the poor.

What is to be done with the rain fed regions till then? There have been several successful small experiments in different parts of both the rain fed and dry land areas of the country. For example, there has been considerable success in some pockets of Maharashtra in conserving water use by the community. This has helped in raising suitable crops and livestock and in creating a viable market system. We should recall how the green revolution took place. Several farmers from the irrigated regions of India were given opportunity to visit other parts of the world.

Table 11 : Projected household demand for food in India at 7% income growth

annual household demand (Million tons)

Commodity	1991	1995	2000	2010	2020
Food grains	168.3	185.1	208.6	266.4	343.0
Milk	48.8	62.0	83.8	153.1	271.0
Edible oil	4.3	5.1	6.3	9.4	13.0
Vegetables	56.0	65.7	80.0	117.2	168.0
Fruits	12.5	16.1	22.2	42.9	81.0
Meat, fish and eggs	3.4	4.4	6.2	12.7	27.0
Sugar	9.6	10.9	12.8	17.3	22.0

Source: TIFAC, Food and Agriculture: Technology Vision 2020.

Our people and farmers are all integrated into one huge market. All concerned need to be educated about another important scientific fact through observations, discussions and mass contacts; i.e., the agro- ecological considerations. Each state should concentrate on agricultural products most suited to its agro climatic conditions. Suitable marketing and transportation systems can be evolved to facilitate the exchange of commodities. There is need for multiproloned action. Merely having better seeds or better irrigation will not suffice. The task involved today is much more complex for raising the organic agriculture than they were during the green revolution.

Vision

It is difficult to express this vision and action in few words. However, to focus on crucial issues, we have attempted to list a few important items below:

- India to aim to be a major player in the world in the organic farming and a leading exporter of grains and other agricultural products.
- Eastern India to become a major producer of organic wheat.
- Rice- producing areas to use organic seeds on a large scale.
- Central India to be made a centre of organic vegetables, fruits, pulses and coarse grains.
- More emphasis on tuberous crops.
- Water as a national resources and water management as the key to agricultural prosperity.
- Steps to educate farmers about the technologies related to the organic farming, if needed by providing them the opportunity to travel, and use of specific technologies to facilitate interaction and encourage farmers to ask questions and share experiences.

REFERENCES AND CITATIONS

Abdul Kalam, A. P. J and Sivathanu Pillai, A. 2004. Envisioning an Empowered Nation. Tata McGraw-Hill Company Limited New Delhi, India .255pp.

All India Co-ordinated Research Project on Microbiological decomposition and Recycling of Farm and City wastes. Progress Report, Indian Institute of Soil Science, Bhopal, India p.85.

Amar, B and Cisse, L; The Key Role of Phosphate Fertilizer for Achieving balanced Fertilization, Food Security and Quality in Sustainable Agriculture. World Phosphate Institute, 3, Rue Albdelkader Al Mazini, Casablanca 20001-Morocco.

APEDA, 2008.

Badgley, C. 2006. Organic Agriculture and the Global Food Supply, Description.

Bedekar, V. V (ed.). 1993. Agriculture in Ancient India. Itihas Patrika Prakashan, Thane 400602, India. 212 pp.

Bhat, M. R. 1981. Varahamihira's Brhat Samhita. Part-1. Motilal Banarsidas, Delhi 110 007, India.pp 527-535 (Reprint,1992).

Bhattacharya, P. 2004. Organic Food Production in India Status, Strategy and Scope. Agribios (India) Jodhpur 01-182 (2004).

Chhonkar, P. K. J. 2003. *Soil. Sci. Soc. India*, Vol.51, No.4, 365-377.

Control Union World Group. EEC Regulation No. 2092/91

Definition of Organic Agriculture. IFOAM. http:// wwwdefinations/doa/index html. Retrieved 2008-09-30.

Dimitri, C; Oberholtzer, I. 2006. EU and US Organic Markets Face Strong Demand Under Different Policies.

Directorate General for Agriculture and Rural Development of the European Commission What is Organic farming.

FAO, 2006

Gillman, J. 2008. The Truth About Organic Farming. Timber Press.

Hanson, J.C; Lichtenberg, E and Peters, S.E. 1997. Organic verses conventional grain production in the Mid–Atlantic. An Economic and Farming System Overview. *Ann. J. Alternative Agric* 12.2.

http:// www.agricultureinformation.com. 10.2.2008.

http:// www.ifoam.org/growting organic/ definations/doa/index.html.retrieved 2008-09-30.

http:// www.tribuneindia.com/2004/20040524/agro.htm#2

http://www.agriculture information.com

IFAOM. 2005. The IFAOM Norms.

IFOAM. 2004. The World of Organic Agriculture Statistics and Emerging Trends. (Helga Willer and Minou Yussefi Eds). 1-167.

India's Booming Organic Food Bazaar. 2010. Rediff.com.

International Federation of Organic Agriculture Movements. IFOAM. 2008. Basic Standards for Organic Production and Processing, IFOAM Publication, Tholey-Tholey, Germany, 1998.

IPFS, Farming Group Org. Integrated Polyculture Farming System.

Kortbech-Oleson, R. 2000. Export Opportunities of Organic Food from Developing Countries. In world Organics, Agro Europe (London) Ltd. London, UK.

Lampkin, Nicholas; Foster, S; S. Padel and P. Midmore. 1999. The Policy and Regulatory Environment for Organic Farming in Europe In: Organic farming in Europe: Economics and Policy. Vol-2 Dabbert, Haring, Zanoli Eds. London. Zed Book.

Maeder, P. 2008. Soil Fertility and Biodiversity in Organic Farming. *Science*. 296.

Magkos, F. 2003. Organic Food: Nutritious Food or Food for thought? A Review of the evidence. *International Journal of Food Science and Nutrition*. 54 (5): 357-371.doi:10. 1080/ 09637480120092071.PMID12907407.

Mahapatra, B.S; Ramasubramanian, T; Chowdhury, H. 2009. Organic Farming for Sustainable Agriculture: Global and Indian Perspective, *Indian Journal of Agronomy*, 54(2).

Marshall, G. 1991. *Review of Marketing and Agricultural Economics*. 59 (3):283-296.

Mohan Singh. 2003. Organic Farming- Prospects in Indian Agriculture. Souvenir of 68th Annual Convention of Indian Society of Soil Science. CSAU & T, Kanpur, 4-8 Nov. pp 52-60.

Narlikar, J.V. 2003. The Scientific Edge from Vedic to Modern Times. Penguin Books, New Delhi, India. 224pp.

National Organic Programme Regulations.

Nene, Y.L and Sadhale, Nalini.1997. Agriculture and Biology in Rig-Veda. *Asian Agric-History* 1: 177-190.

Offer Mann, F and Neiberg, H. 1999. Economic Performance of Organic farms in Europe. Organic farming in Europe: Economics and Policy 5, University of Hohenheim, Hohenheim, 1999.

Organic Agriculture and Poverty Reduction in Asia: Overview of Markets and Marketing. SOEL-Survey (February 2004) of 1998-2002 Data; Graph: Yussefi, SOEL.

Padel, S and Lampkin, N. H. 1994. Farm level performance of Organic Farming Systems. An Overview. In Economics of Organic Farming (Eds Lampkin, N.H and Padel, S). CAB Walling Ford, U.K.1994, pp 201-219.

Palaniappan, S. P and Annadurai, K. 1999. Organic Farming-Theory and Practice. Scientific Publishers (India) 1-257.

Paull, John 2007. "China's Organic Revolution", *Journal of Organic Systems*. 2(1):1-11.

Peterson, C; Drink Water, I and Wagoner, P. 1999. The Rodale Institute Farming System Trial: The first 15 years. The Rodale Institute Kutztown, PA 1999, p40 www. rodale institute org.

Randhawa, M. S.1980. A History of Agriculture in India. Vol.1. Indian council of Agricultural Research, New Delhi, India. 541 pp.

Regan old J. P, Palmer A. S. 1993. Lockhart, I. C; and Macgregor, A.N. *Science*, 260: 344-349.

Regan old, J.P; Glover, J.D; Andrews, P. K and Hinman, H.R. 2001. Sustainability of Three Apple Production Systems. *Nature*, 410, 926-930.

Sadhale, Nalini. (TR). 1999. Krishi Parashara (Agriculture by Parashara). Agric-History Bulletin No. 2. Asian Agric-Foundation, Secunderabad 500 009, India. 104pp.

Sensarma, P. 1989. Plants in the Indian Puranas. Naya prakash, Calcutta 700 006, India, 193pp.

Shamasastry, R.1961. Kautilya's Arthashastra. Seventh Edition. Mysore Printing and Publishing House, Mysore, India 482 pp. (First Edition published in 1915).

Sharma, G. S. 1991. Rigved. (In Hindi) Sanskrit Sahitya Prakashan, New Delhi, India.769pp.

Strochlic, R; Sierra, I. 2007. Conventional, Mixed and De-registered Organic Farmers: Entry Barriers and Reasons for Exiting Organic production in California Institute for Rural Studies.

Swaminathan M.S. (Ed) 1993. Wheat Revolution-a Dialogue. Macmillan India Limited, Chennai, India 164 pp.

Tampoli, P. M. and Nene, Y. L. Science in India with special reference to Agriculture.

Tandon, H.L.S. 2002. *Fertilizer Guide*. 21.

TIFAC, Food and Agriculture: Technology Vision 2020.

USDA NOP Program Standards. 2008.

Vijnan Bharti. 2002. Science and Technology in Ancient India. Mumbai, India.118pp.

Watson C A, Atkinson D, Gosling, P, Jackson, I R, Rayns F W. 2002. "Managing Soil Fertility in Organic Farming Systems" Soil use and Management 18:239-247.doi:10.1111/ j.1475- 1743.2002.tb00265.x.http.//www3.interscience.wiley.com/journal/ 119192119/abstract.retrieved 2009-05-29.

Willer, Helga and Kilcher Lukas (Eds). 2009. The World of (Eds). 2009. The World of Organic Agriculture, Statistics and Energy Trends 2009. IFOAM, Bonn; FiBL, Frick; ITC, Geneva.

❑❑❑

Chapter 6

Exploitation of Traditional Knowledge in the Management of Crop Diseases under Organic Agriculture in India

SHAMARAO JAHAGIRDAR*, BENAGI V.I.*, SANGSHETTY G. BALEKUNDE** AND NAGARAJU, P.*

**Department of Plant Pathology, University of Agricultural Sciences, Dharwad-580 005, India*
***Dept. of Agronomy (Plant Molecular Breeding Lab), Chungnam National University, Daejeon, South Korea*

INTRODUCTION

More than 200 million farmers and farm workers are the backbone of Indian agriculture. The establishment of an agrarian economy is the mainstay of reforms process in Indian agriculture sector which provides food security to more than one billion people and also provides raw materials for industry and exports. Organic agriculture is a reward enterprise for both farming community and natural environment under changing global climate scenario. Sustainable agriculture promotes technically sound, economically viable, eco-friendly, non-degrading and socially acceptable use of country's natural resources such as land, water and genetic resources. The organic has got multi fold dynamism in rational utilization and conservation of abundant water, conjunctive use of surface and ground water, special care for water quality, tackle the problem of ground water depletion, proper on-farm management of water resources for irrigation purpose. The history of traditional knowledge of agriculture, dates back to Vrikhayurveda where in there was mention of ailments curing in plants by using locally available resources. This gives hint on management of crop diseases by using indigenous technology knowledge. India has a very strong base in organic agriculture and management of crop diseases. The major thrust of organic agriculture is on pooling, distilling and evaluating traditional wisdom and harnessing it for sustainable growth. In this chapter all such traditional knowledge will be explained with scientific background and research findings in some selected crop diseases.

RENEWED INTEREST TOWARDS ORGANIC AGRICULTURE IN INDIA

The Scientific cultivation methods in organic agriculture started with environment and health-conscious people of the developed world. However, organic agriculture is for securing a place on international markets, export promotion, economic self-reliance, finding alternatives to decreased access to agricultural inputs, natural resource conservation, food self-sufficiency, and rural and wider social development.

In India, especially in low-input traditional systems, properly managed organic agriculture systems could increase agricultural productivity and restore the natural resources base. The main aim of India's policy regarding organic agriculture is income generation through the promotion of certified organic food. For many, especially policy-makers, organic agriculture is a set of strict rules and complicated practices that allow marketing of certified food products. In developing countries, policies for organic agriculture seek to earn foreign exchange for other development needs through exports. In fact, declining government budgets have forced many developing nations to re-structure their agriculture sector. Agricultural institutes and extension agencies have to provide more market-oriented services. Liberalization and privatization policies have paved the way for a greater role for entrepreneurs and producers. These trends act as a catalyst for private initiatives in the field of organic agriculture. When a critical mass of practitioners is gathered for organic agriculture, government formulates policies to support the marketing of certified organic products. The incentive of such policies is therefore economic, whether for tapping lucrative markets, securing a place in world trade and/or counter-balancing withdrawal of government support to agricultural inputs or other services.

However, a large number of small-scale subsistence farmers in India produce simply for consumption, do not participate in the market and are left behind due to globalization. The major challenge is to establish organic agricultural policies that combine income generation and improved domestic food production. The latter involves raising productivity of poorly endowed areas by maximizing the use of local resources. Such policies would better respond to self-reliance, local food needs and health of resource-poor farmers. In India, the interest in organic agriculture is growing because it requires less financial inputs and places more reliance on the available natural and human resources. Studies undertook to date seem to indicate that organic agriculture offers a comparative advantage in areas with less rainfall and relatively low natural soil fertility levels. In fact, agricultural labour realizes a good return, which is very important where paid labour is almost non- existent.

Organic agriculture does not need costly investments in irrigation, energy and external inputs but rather substantial investments in capacity building through research and training. Pro-poor organic agricultural policies have the potential to improve local food security, especially in marginal areas.

PLANT DISEASE MANAGEMENT IN ORGANIC AGRICULTURE

Plant diseases are the major constraints of yield reduction in various crop plants. There are about 1000 diseases which affect economic crop plants causing significant damages. Majority of plant diseases are caused by fungi followed by viruses, bacteria, nematodes, phytoplasms and a number of other plant pathogens. Imbalance of nutrients and extreme variation in environmental factors also leads to serious plant diseases. Integrated management of these diseases needs a synergy with the natural environment.Tradional agricultural practices were tailor made to maintain this synergy and thus, forms the basis for plant disease management. Organic agriculture is usually associated with pre-industrial peasant agriculture. Indigenous knowledge is the largest single knowledge source not yet fully exploited.

TRADITIONAL KNOWLEDGE IN MANAGEMENT OF PLANT DISEASES

India has a treasure of Indigenous knowledge concerning plant health developed and documented several centuries ago. The three major ancient texts *inter alia* are

(1) Varahamitra, 505AD, Brihat samhita, Vrikshayurveda Part 1, Chapter 55(edited by M Ramakrishna Bhat, Motilal Banarasidas, Bangalore, India, 1950)

(2) Chavvdaraya, 1025 A.D Lokopakaran, Vrikshayurveda, Chapter 6(edited by H.Shesha Iyenger, Government Central manuscripts Library, Madras, India, 1950).

(3) Sarangahara, 1363 A.D.Vrikshayurveda (edited by S.K.Ramachnadra Rao, Kalpatharu Research Academy, PO Box, 1857, Bangalore, 1993, India).

The most detailed of all the Vrikhsayurveda work, seems to that of Chaundaraya, 1025 A.D.It is however to be noted with caution that development of this body of knowledge needs to be viewed in pest disease scenario prevalent 10 to 15 centuries ago, during which many of the current pests/diseases may not have existed.However, the methods employed have characteristics such as (1) Multi pronged attack on the pest/disease. (2) Improving plant health thereby increasing resistance capability. (3) Enriching soil with buffering of useful microbial activity and (4) Broad spectrum effects on pest/disease which are desirable.

Certain details as under, though these may not be necessarily practicable at this point of time. 1. A few indigenous methods of plant protection as outlined in ancient text. 2. The documentation of research endeavour in a few indigenous knowledge practices.3.Local practices adopted for plant disease management.4.Policy implications

1. A Few Indigenous Methods of Plant Protection in Ancient Texts

Esteemed authors of ancient texts make a mention of their painstaking efforts while observing and researching on indigenous methods that have been documented. In this paper mainly methods of disease control are highlighted.

The following methods are drawn from Lokopakara, chapter VI-Vrikshayurveda.

a) Disease Avoidance (as in 12th stanza): The incense of *Embelia ribes* Burm (Vayuvilanga) fam : Myrsinanceae). *Commiphorawightii* (Mahishakshi) fam : Burseraceae), fishmeat, *Curamalonga* (asisna) fam : Brassicacea) flowers of *Terminaliatomentosa* (methi) fam: Comdretaceae) will provide resistance to plant diseases. If growing tips of plants dry up or set broken or get whitened such plants are inferred to have been affected by disease. In the affected plant, warm ghee has to be smeared and black soil applied to combat the disease.

The following methods are drawn from Brihat Samihitha, Part 55, Chapter VI, Vrikshayuverda.

b) Seed Treatment

The seeds need to be soaked in cow's milk for ten days. Every day, they have to taken out with palms greased with cow ghee.Later, they are rolled in cow dung mixed with flesh of deer and hog. Thus treated soils are planted in soil treated with sesamum.The seeds sown are to be sprinkled with milk and water.

c) Disease Treatment

The ulcers of affected trees be removed with knife. A paste made using *Embelia ribes* (Vayurlanga) fam:Myrisinaceae), cow's ghee and silt be applied to the affected parts and later be sprinkled with water and milk

The following methods are drawn from Vrikshayurveda of Saranadhara.

The pathological conditions of wind(Vata), bile(Pitta) and phlegon(Kapha) which are responsible for diseases in human beings are causes of diseases in plants also. When trees are affected by pests, the affected parts are to be removed. The diseases due to vata(wind) are overcome by application of clarified butter and fresh juice, that due to bile is overcome with substances that are cold mixed with water and that due to kapha(phlera) are cured with substances which are acidic mixed with hot water or pungent and bitter substances. Plants attacked by pests are treated with mixture of fresh cow urine, clarified butter, *Embelia ribes* Burm, mustard and sesamum applied to trunk and then are to be watered with milk and water. Trees which are attacked by any type of pests are to be treated with the paste from bark of *Pongamia glabra* (Honge) (fam:Poaceace), *Cassia fistula* (Kakkagida) (fam : Caeslapiniaceae), *Sapindus laurifolius* (arishta) (fam : Sapindaceae), *Alstomiascholaris* (Saptaparni) (fam : Apocynaceae), *Embelia ribes* Burm (Vayuvilanga) (fam : Mysinaceace) and *Cyperus esculentus* (fam : Mysinaceae) and *Cyperus esculentus tengamurte* (fam : Cyperaceae) in cow's urine. Exudation from trees can be stopped with application of bark paste of *Setaria italica* (Priyangu/Navane)(fam:Poaceace) and *Termindelia arjuna* (arjuna) (fam:Combretaceae) in boiled milk.Surapala in Vriksahvyrveda suggested measures of treatment of plant diseases. Trees that grow too close and touch each other do not yield adequate fruits owing to their roots entwining and injuring each other, so recommended appropriate plant to plant distance as applicable to individual species should be follwed. Diseases like the scoring of leaves, arrest of the growth of leaves, drying up of branches and excessive exudation of sap have recommended remedies like clearing of the affected part and then pestering them with sriangdhara and ghee and nurturing the well mixed with milk. Cessation of fruits bearing (i.e.sterility) by application of kulattha, masammurdga seasmum along with honey. Traditional farming practices in past have provided effective and sustainable management of crop diseases. Most of present day susceptibility of cultivars have been avoided from right technique and planting materials. They practised carefully seed soaking, crop rotation and avoided planting at full moon or sun had a halo. Most of traditional farmers strictly adhere to plant and crop artchitecture.Cultural control practices such as fresh burning, adjusting crop sowing depth or time of planting, fallowing, flooding, mulching, cropping with zero tillage, planting in raised beds, sanitation and tillage.

OUR RESEARCH EFFORTS IN MANAGING CROP DISEASES BY ITK MEASURES

In validation of Indigenous Technology Knowledge, UAS, Bangalore is the pioneer research institute in the country which first time took research efforts in plant protection. In general management of soil borne plant pathogens is very difficult particularly Fusarium group which is soil inhabitant and remains in soil for more than 50 years in soil.Hence, it is difficult to manage this pathogen by any methods including chemical methods.

The cow milk, curd, ghee, cowdung and cow urine have been used individually used as for curing many ailments as described in ancient text. It is to know that cow ghee and curd contain certain living entities and antimicrobial substances. The Panchagavya is the product of five cow products such as milk, curd, ghee, dung and urine. In tradional Hindu family it is also taken as Panchamratha in little quantity for purification both external and internal environmental of system. The innovative research on use of Modified Panchagavya Mixture (MPG-3) was carried out on three soil borne diseases like Fusarium wilt of tomato and Banana, also foot rot of black pepper. The traditional panchagavya was modified by adding yeast and common salt and three formulations were tested. The component three MPG-3 was most effective in managing all these plant diseases which includes 2ml of ghee, 5ml of curds, 5ml of milk, 40g dung and 48ml of urine mixed with 2g yeast and 2g salt for 100ml preparation. These components were mixed by adding one after the other in plastic container and kept for fermentation for seven to 10 days by closing of plastic container. The addition of salt is to reflect Jim Martin's living water promoting microbial activity which is further augmented with addition of yeast. The fermented preparation was diluted ten times with water and filtered through two layers of muslin cloth to obtain clear filtrate. The filtrate was used in different delivery methods of seedling dip for 30 minutes and soil drenching for the pre-infested soil with the pathogen in the investigations.

i) Management of panama disease of banana—In case of Panama disease of banana, MPG-3 was used at 10^1 dilution along with different bioagents like *Trochoderma viride*(0.25%).*Pseudomonas fluorescens* (1 hr dip, 10^8 cells/ml).*Bacillus subtilis* (1 hr dip, 10^6 cells/ml).The MPG-3 gave better influence on plant height, number of leaves, maximum root length and pseudo stem girth etc.There was reduction in Fusarium population in MPG-3 provided encouraging results compared with seedling dip.The population of *Fusarium oxysporum* f.sp.*cubense* declined significantly to 11.8×10^4 cfu/g after 150 days of planting. These results indicate the promise shown by MPG-3 in eco-friendly and cost effective management of Fusarium wilt (Shamarao Jahagirdar, 1994, Shamarao Jahagirdar *et al* ., 2000).

ii) Management of foot rot of black pepper—Developed and standardized effective IDM package: Soil application of *T.viride* (75g/pt)+spraying with metalaxyl (1.25g/lt)+Akomin(4ml/lt) or MPG 3 (10^1) for the management of foot rot of black pepper (Shamarao Jahagirdar , 1995, Shamarao Jahagirdar et *al*., 2000).

iii) Management of damping off of tomato in nursery and main field—The research work carried out by UAS, Bangalore clearly demonstrated the role of MPG-3 as PGPR component and ISR activity against Fusarium wilt of tomato (Baskar Padmodaya, 1994).

iv) Management of tobacco mosaic virus through organics—Tobacco Mosaic Virus (TMV) is the major stumbling block for successful cultivation of bidi tobacco in Nipani area. Identification of resistant source against such systemic biotic infection is a challenging task for Plant Pathologists and Plant Breeders. In order to give a boost to ruling cultivators which are susceptible for TMV, Application of Viroson @2% (27.7 Percent Disease Incidence) followed by Bougainvillea leaf extract @ 5%(30.2% incidence) and neem 1500ppm(31.8%) incidence. Among ITK measures, Panchaghavya @5%(37.7%) followed by cow urine@10% (37.8%)incidence. The untreated check recorded maximum incidence of 56.5 % incidence.

There was no significant difference among the treatments with respect to growth parameters. However, higher plant height, leaf length and leaf breadth were recorded in Viroson, Neem 1500ppm and Cow urine application indicating role induced systemic resistance. Maximum cured leaf yield (1206kg/ha) was recorded in cow urine@10% followed by Viroson@2%(1157 Kg/ha). Among quality parameters nicotine % ranged from 2.66 to 4.16 with maximum (4.16) in neem leaf extract followed by 3.77 % in Butter milk@5%.The reducing sugar ranged from 5.63 to 10.14% with maximum (10.14%) in Neem @1500ppm followed by 9.78% in Cow urine@10%. The Chloride % was within the limit of <1 except butter milk (1.07%).Thus, the investigations opened a new window of opportunity in managing TMV infections through ITK measures enhancing both leaf yield and quality parameters in bidi tobacco(Shamarao Jahagiradar et al., 2008).

v) Management of Asian soybean rust in India—The Asian Soybean rust, *Phakopsora pachyrhizi* Syd is the economically important disease, which causes significant yield loss in India. Lack of resistant cultivars, growing concern over use of chemical pesticides and increasing area under organic soybean cultivation has lead to exploitation of Indigenous Technology Knowledge in the management of Asian Soybean Rust. The pooled analysis over two years revealed that among the ITK measures application of Cow urine@10%+*Pongamia pinnata* oil @ 0.5% recorded minimum Percent Disease Index (PDI) of 37.9 followed by Cow urine@10% alone (40.3). The chemical elicitors like MnS04, Muti-k or plant based extracts like *A.vesica.Pongamia pinnata* oil and bioagent like *Trichoderma harzianum* along with cow urine be used in developing Integrated Disease Management strategies against Asian Soybean rust in India which will help in reducing the chemical pesticides in long term sustainable management. The present findings drawn the first line of research on utilization of Indigenous Technology Knowledge in managing rust and enhancing both yield and quality parameters of soybean in India.(Shamarao Jahagirdar et al., 2009).

vi) Eco-friendly strategies in the integrated management of root knot nematode in bidi tobacco-2003 to 2007—Preamble: Different types of tobacco are being cultivated in India under different agro climatic conditions. Among these, the Nipani area of Belgaum District of Karnataka in India is known for production of quality bidi tobacco. Monocropping of tobacco in this region has resulted in building up populations of *Meloidogyne incognita* (Kofoid and White) Chitwood. The disease has become a constraint for tobacco cultivation in the area. It infects at any stage of the crop and causes considerable loss in quality and yield of tobacco. Bidi tobacco is bread and butter of Nipani farmers of Northern Karnataka. Our four year research efforts finally gave solid recommendation as poultry manure (1 t/ha) mixed with Carbofuran 3G (5 kg/ha) was the most effective, suitable, eco-friendly and economically viable strategy for the management practice of root-knot disease of bidi tobacco. This has significantly reduced the use of carbofuron in the area.(Shamarao Jahagirdar & Hundekar, 2007a:2007b and 2008).

SOME OF THE CLASSICAL EXAMPLES OF LOCAL PRACTICES ADOPTED FOR PLANT DISEASE MANAGEMENT ARE

- Shade regulation for coffee leaf rust and blister light of tea.
- Growing wind breakers like silver oak, casurina, jack etc to avoid sun scorching of young shoots of plantation crops.
- Tieing of areca nut seedlings with coconut and areca nut fronds to protect them from western sun scorching.

- Pasting of lime on areca trees to avoid ill effects due to sun scorching.
- Watering nursery beds in early morning for higher seedling vigour and stand particularly followed in Chilli and Brinjal.
- Burning nursery with leaf litter and farm waste to overcome certain soil borne pathogens.
- Raised beds, fields and ridges used to manage some soil borne pathogens. In Mexico, raised beds are called as Chinampas or floating garden were used to control *Pythium, Phytophthora* and other soil borne pathogens.
- Collection and burning of stubbles in the field to over come the problem of soil borne pathogens.
- Flooding with water to over come the problems of soil borne pathogens by creating anaerobic conditions. In our studies flooding for 85 to 100 days brought down significantly the *Fusarium oxysporium* f.sp.*cubense* population causing panama disease of banana.
- Earthing up to over come the problem of pythium damping off in nursery in brinjal and tomato.
- Kotte tieing for areca bunches to over come problem of koleroga of areca nut.
- Planting across the wind direction helps to manage some airborne diseases.
- Mixed cropping of jowar with tur to prevent the movement of mites which transmit sterility mosaic of pigeon pea and to minimise tur wilt.
- Manipulation of planting time/ sowing to over come problems of foliar diseases eg.Tikka disease of ground nut and anthracnose of chilli
- Summer ploughing to reduce the problem of nematode infestation and soil borne pathogens.
- Magi cultivation: Crop rotation with legumes, cereals and millets to overcome the problem of soil borne plant pathogens.
- Mulching with green manure in paddy to over come problem of soil borne pathogens.
- Salt water treatment for wheat to over come the problem of seed borne diseases eg.Bunt and seed gall in wheat.
- In North Costarica traditional farmers in many areas used a system called tapaga meaning Covered beans and combination of mulch and beans effectively prevented bean blight.
- Benching of maize earheads in Mexico to over come the seed borne infestation due to fungal pathogens.
- Tieing of paddy thread prepared out of hay near the crown region of coconut and placement of 1kg of salt to over come the problem of stem bleeding in areacnut. (Shamarao Jahagirdar *et al.*, 2003).

ROLE OF BIOLOGICAL CONTROL IN ORGANIC AGRICULTURE

Plant diseases are managed through different practices employing different agents but management through bioagents is called as biological control of plant diseases. Although Plant pathologists had used the term biological control agent in the early 1900s, there was no attempt to formally define the term biological control. Baker and Cook (1974) defined biological control as "reduction of inoculum density or disease producing activities of a pathogen or parasite in its active or dormant stage by one or more organisms accomplished naturally or through manipulation of environment, host or antagonist, or mass introduction of one or more antagonists". Further Cook and Baker shortened this definition to: " Biological control is the reduction of the amount of inoculum or disease producing activity of a pathogen accomplished by or through one or more organisms other than man".

Mukhopadhyay (1987) highlighted that "Biological control of soil borne plant pathogens by *Trichoderma* spp and other bioagents as a vital area of plant pathological research all over the world these days".Biological plant protection is too important component in the ecofriendly management of plant diseases all over the globe. It is now widely accepted that biological control of crop diseases is a distinct possibility for the future and can successfully be exploited within the frame work of **Integrated Pest/Disease Management System**.

Why we are giving importance to biological control of plant diseases?

Following are some of the points presently supporting management of plant diseases through bioagents.

1) It avoides environmental pollution of soil, air and water as is being experienced in chemical control.
2) It avoides the residue toxicity of crop products when consumed but it is very much experienced in fungicidal control.
3) It avoides adverse effect on beneficial microorganisms including antagonist in the soil where as chemical are lethal.
4) It is lees expensive compared to chemical control as they are more expensive.
5) Continuous use of bioagents avoides development of resistant strains while continuous use of systemic and specific fungicides has resulted in development new resistant strains thus making chemical ineffective.
6) Bioagents application is usually once and doesn't need repeated applications where as chemicals need to be repeatedly applied as they loose effectiveness after some time.
7) Bioagents are more effective especially for soil borne diseases, as fungicides may not reach pathogen site.
8) In the absence of satisfactory control measures in virus diseases, biological control is presently satisfactory.
9) It aims at risk free management of plant diseases where in chemical control are risky including phytotoxcity.
10) It becomes part of modern large-scale agriculture and helps in increasing crop production within existing resources maintaining biological balance.

Advances in the Approaches to Biological Control

The fundamental approaches are four: Biological control of inoculum, which includes, a. Destruction of propagules or biomass of pathogen by hyper parasites, hyperpathogens or predators. b. Prevention of inoculum formation. c. Weakening or displacement of pathogen in infested residues (food base) by antagonists. d. Reduction of vigor of virulence or pathogen by agents such as mycoviruses or hypo virulence determinate,

2.Biological protection agents infection: It is achieved by: i) Protection of planting material. ii) Protection of roots with biological seed treatment. iii) Biological protection of foliage and flowers. iv) Inoculation of pruning wounds with antagonists.

Mechanisms of Biological Control

Antagonism includes antibiosis, competition and mycoparasitism and the mechanism of biological control of plant diseases operates through one or both or all of these together or singly. In addition, the mycorrhizae, plant growth promoting rhizobacteria(PGPR), cross protection and induced resistance are also operating during the biological control process.

Antibiosis—Antibiosis is the inhibition of pathogen by the metabolic (antibiotic) product or products of the antagonist. The antagonist releases antibiotics or other metabolic products (enzymes), which are harmful to the pathogen and inhibit its growth.

Antagonist	Antibiotic Produce
Agrobacterium radiobacter	Agrocin 84
Glicladium virens	Gliotoxin, Viridin, Gliovirin
Trichoderma viride	Gliotoxins, Dermadine, Viridin, Trichodermin
Bacillus subtilis	Bulbiformin
Pseudomonas fluorescence	Phenazines, Pryoluteorin, Pyrrolnitrin
Streptomycin	Aureofungin, kasugamycin, Streptomycin, Cyclohexamides
Penicillin spp.	Griseofulin, penicillin

Competition—It is the endeavor of two or more micro organisms to gain the measure each wants from supply of substrate in the specific form and under specific conditions in which that substrate supply is not sufficient for both. In essence the competition is for nutrients (high energy carbohydrates and nitrogen), and also for space and oxygen but not for temperature, PH and water potential. The antagonists grow very fast and utilize all the food and occupy the space and thus make the pathogen weak. Heterotrophic rhizobacteria like fluorescent pseudomonas compete for iron with plant pathogens, use very efficiently iron and produce siderophores (microbial iron transport agents) which complex the iron and thus affects them adversely.

Mycoparasitism and Predation—Mycoparasitism (= Hyperparasitism) is defined as parasitism of one fungus by another. Several necrotrophic mycoparasitism have potential bio control agents. The mechanism of mycoparasitism includes different kinds of interactions like coiling of hypae around the pathogen, penetration, production of hastoria and lyses of hypae. Recently, it is also postulated that necrotrophic parasites kill susceptible by the action of toxins, antibiotics and or enzymes.

Induced Systemic Resistance—In the later part 1990s.the research on Plant Growth Promoting Rhizobacteria(PGPR) and Induced Systemic Resistance (ISR) clearly gave a hint on the role of useful microbes in imparting resistance to plants for specific group of pathogen. Free-living root colonizing bacteria (rhizobacteria) have been studied for the past century as possible inoculants for increasing plant productivity and controlling microbial pathogens. .Soil or seed applications with plant growth-promoting rhizobacteria (PGPR) have been used to enhance the growth of several crops (Glick, 1995). as well as to suppress the growth of plant pathogens. PGPR that colonize root systems through seed applications and protect plants from foliar diseases include *Pseudomonas spp. Bacillus spp. Paebacillus spp., and Serratia* sp. The mechanisms for plant growth promotion and induced systemic resistance (ISR) by PGPR have been extensively studied in the past decade. There are several determinants for mechanisms of growth promotion that include bacterial synthesis of the plant hormones (indole-3- acetic acid (IAA), cytokinin, and gibberellin), breakdown of plant-produced ethylene by bacterial production of 1- aminocyclopropane-1-carboxylate (ACC) deaminase, and increased mineral and N availability in the soil. Recently, the phenomenon that PGPR elicit plant defense has also been found to lead to a state of ISR in the treated plant. ISR occurs when the plant's defense mechanisms are stimulated and primed to resist infection by pathogens ISR is different from systemic acquired resistance (SAR) that triggers systemically plant defense response following hypersensitive response after inoculation of plant pathogens . Previous works demonstrated that several bacterial determinants such as siderophores, salicylic acid (SA), and lipopolysaccharides (LPS) contributed to ISR (Choong-Min Ryu, *et al.*, 2005).

Role of Biological Control in the Integrated Disease Management (IDM)

As biocontrol agents alone may not completely and effectively manage the disease, it can be done of the components in the integrated management of diseases. Also under high disease pressure or pathogen population pressure, the biocontrol may be less effective and needs other practices also to completely manage the disease. To site an example, Jahagirdar *et al* (2001) reported the effectiveness of seed treatment with biocontrol agent if integrated with other management practices i.e. sue of moderately resistant cultivar + biocontrol + FYM helped in managing the disease. The effectiveness of biocontrol agents in other crops against soil borne pathogens are presented in table for reference.

Table 5 : Some examples of successful control of various pathogens in field by *Trichoderma*

Pathogen	Crop	Antagonists	Method and rate of application
Sclerotium rolsfii	Tomato, Beans, Ground nut, chick pea	*Trochoderma viride, T. harzianum, T. hamatum*	Furrow 13-160kg.ha.Seed treatment 4g/kg.
Rhizoctonia solani	Ground nut, cucumber, rice	*As above*	As above
Macrophomina phaseolina	Beans Sorghum	*T. harzianum* *T. viride*	Furrow application 13-160kg/ha.Seed treatment 4g/kg of seed.
Fusarium oxysporum	Beans, redgram, cottn, melons, banana	*T. harzianum, T. viride*	Soil 16okg/ha for beans 130-160 kh/ha for other crops. Banana 75g/plant mixed with 2kg of FYM per plant
Botrytis cinerea	Grey rot of grapes	*T. viride*	Aerial spray

(Srikant Kulkarni, 2004)

Keys to the Success of Biological Agent

1. Selection of virulent strain of antagonist.
2. There should be sufficient growth and sporulation on the mass culture media incase of facultative before applying.
3. Advance application to provide enough time for interaction between antagonist and target pathogen
4. The soil temperature, moisture should be optimum for establishment and growth of antagonist.
5. Sufficient organic matter (food base) should be present in the soil for multiplication, survival and activity of the antagonist.
6. Monitoring the population of the target pathogen and antagonist from time to time on selective media.
7. Handinlng, production, storage should be easy and also product should be cheap and available.
8. Integration of biocontrol with tolerant / resistant varieties and/or chemical control may be more feasible.

Future line of research work with respect to biological control

1. Isolation and evaluation of native antagonists, their multiplication and preparation of seed treatment formulations.
2. Mapping up of population dynamics of pathogen and biocontrol agents in various geographical areas of production systems.
3. Application biotechnological techniques, genetic engineering, protoplast fusion technique etc. may be employed in developing efficient strains of antagonists of sites specific, area specific and crop specific would further pay way for faster developments in biological control.

IMPLICATIONS

Plant protection Experts all over the world are beginning to evince keen interest in indigenous methods of plant protection. Of late this knowledge is being made available, research endeavours may be reoriented towards validation of indigenous methods encouraging Integrated Disease Management (IDM) practices. Indigenous knowledge by itself cannot be a sole method of management but integration with other methods would surely pave way for effective management of diseases without affecting natural ecosystem adversely.

REFERENCES

Anonymous., 1964., Agriculture in ancient India.Indian Council of Agricultural Research, New Delhi, 167pp.

Bhaskar Padmodaya.1994. Integrated management of Fusarium wilt of tomato, Ph. D thesis submitted to Dept. of Plant Pathology, UAS, Bangalore.

Chandrakant, M.G., Ramachandra Reddy, M.R., Siddaramaih, A.L., Padmodaya, B, Shamarao Jahagirdar and Savaradhy, N., 1995, Plant protection practices in India on indigenous knowledge. Presented in International symposium on "Indigenous knowledge system at 13th International Plant Protection Congress, 2-7 July, 1995, The Hague, the Netherlands.

Choong-Min Ryu, Mohammed A. Farag1, Paul.W. Paré and Joseph W. Kloepper, 2005, Invisible Signals from the Underground: Bacterial Volatiles Elicit Plant Growth Promotion and Induce Systemic Resistance. *Plant Pathol. J.* 21(1) : 7-12.

Curtis, T., 1992., Jim martin living water theory. Span, Feb., 29-37pp.

Kulkarni, Srikant 2004. *Trichodema*-A potential bioagent of Millennium.p.36-44.Paper Presented in 13th Southern Regional Conference on Microbial Inoculants. "Microbes; Wheels of Organic Farming held at Agril.College, Bijapur from 3-5, December, 2004.110pp.

Sarangadhara, 1963, In:Vrikshyurveda (Ramachandra rao ed.), Kalpatharu Research Academy, PB No;1857, Bangalore.

Shamarao Jahagiradar, Siddaramiah, A.L., and Narayanswmy, H., 1997. Mass multiplication of bio-control agent *Trichoderma viride. Karnataka Journal of Agricultural Sciences*, Vol.11: 233-236.

Shamarao Jahagirdar. 1998. Etiology and Management of Foot rot of Black pepper. Ph. D thesis submitted to Dept. of Plant Pathology, UAS, Bangalore. 110pp.

Shamarao Jahagirdar, Siddaramaih, A.L. and Chandrappa, H.M.2000. Eco friendly integrated management of foot rot of black pepper (Piper nigrum L.), *Mysore Journal of Agricultural Sciences*, 34: 47-54.

Shamarao Jahagiradar, Siddaramiah, A.L. and Ramaswamy, G.R., 2000. Influence of Biocontrol agents and MPG-3 on Fusarium oxysporum f.sp. cubense as incitant of panama disease of banana. *Plant Disease Research* Vol.15: 241-243

Shamarao Jahagirdar, Patil M.S., and Indira, S. 2001, Biological control of charcoal rot of sorghum caused by *Macrophomina phaseolina_Agril. Science Digest* 153-156.

Shamarao Jahagirdar and Siddaramaiah A.L.2003, Role of *Pseudomonas fluorescens* in the management of *fusarium* wilt of banana. Paper presented inIn:6th International PGPR workshop held at IISR, Calicut from 5-10, Oct.2003. 630-634pp.

Shamarao Jahagirdar, Biradar, B.D. and Nageshwar Rao, T.G. 2003. Application of *Pseudomonas fluorescens* as low cost eco-friendly technology for management of charcoal rot of *rabi* sorghum caused by *Macrophomina phaseolina* in Northern Karnataka. Paper Presented in 6th International PGPR workshop held at IISR, Calicut from 5-10, Oct.2003.634-637pp.

Shamarao Jahagirdar and Siddaramaiah A.L, 2003. Application of *Pseudomonas fluorescens* in the management of foot rot of black pepper in Karnataka. Presented in 6th International PGPR workshop held at IISR, Calicut from 5-10, and Oct.2003.637-642pp.

Shamarao Jahagirdar, Kulkarni, Srikant and Nageshwar, T.G., 2003. Present status and future research needs on the management of charcoal rot of sorghum. P.15 In National symposium on Recent Advances in the Diagnosis and management of Plant Diseases for meeting Global challenges held at Dept. of Plant Pathology, UAS, Dharwad from 18-20, December, 2003.118pp.

Shamarao Jahagirdar, Ravikumar, M.R. and Siddaramaiah, A.L. 2003. Traditional methods in management of plant diseases *Agricultural Review*, 24 (2): 142-146.

Shamarao Jahagirdar and Hundekar, A.R., 2007a. Ecofriendly strategies in the integrated management of root knot nematode of bidi tobacco P.181 182.In Proceedings of Third Asian Conference on Plant Pathology held at Gadjah Mada University, Indonesia from 20-24, August, 2007.344pp.

Shamarao Jahagirdar and Hundekar, A R 2007b. Eco-friendly strategies in the integrated management of root knot nematode in bidi tobacco-2003 to 2004, (Part I). *Indian Journal of Plant Protection*, 35(2):329-330(IF:3.5)

Shamarao Jahagirdar, Kajjidoni, S.T, Matiwade, P.S and Devappa, V. 2008. Management of tobacco mosaic virus in bidi tobacco in Karnataka through organics. Proceedings of National symposium on Plant Protection held at UAS, Bangalore from 4-6, December, 2008

Shamarao Jahagirdar and Hundekar, A.R. 2008. Eco-friendly strategies in the integrated management of root-knot nematode of bidi tobacco, 2003-2006. Plant Disease Management Reports Vol 2 (APS, Online Journal)

Shamarao Jahagirdar, Patil, P.V. and Basavaraja, G.T. 2009. Role of Indigenous technology Knowledge in the management of Asian Soybean rust caused by *Phakopsora pachyrhizi* in India.In the proceedings of World Soybean Research Conference VIII held at Beijing, China from 10-15, August, 2009.

❑❑❑

Chapter **7**

Plant Growth-Promoting Rhizobacteria for Biocontrol of Plant Diseases

B.K. Mishra, S.K. Khandelwal and A. Joshi

Department of Molecular Biology & Biotechnology, Rajasthan College of Agriculture, Maharana Pratap University of Agriculture & Technology, Udaipur

As agricultural production intensified over the past few decades, producers became more and more dependent on agrochemicals as a relatively reliable method of crop protection helping with economic stability of their operations. However, increasing use of chemical inputs causes several negative effects, i.e., development of pathogen resistance to the applied agents and their non-target environmental impacts. Furthermore, the growing cost of pesticides, particularly in less-affluent regions of the world, and consumer demand for pesticide-free food has led to a search for substitutes for these products. There are also a number of fastidious diseases for which chemical solutions are few, ineffective, or nonexistent. Pathogenic microorganisms affecting plant health are a major and chronic threat to food production and ecosystem stability worldwide. Biological control is thus being considered as an alternative or a supplemental way of reducing the use of chemicals in agriculture.

Plant growth-promoting rhizobacteria (PGPR) are associated with many plant species and are commonly present in many environments. The most widely studied group of PGPR is colonizing the root surfaces and the closely adhering soil interface, the rhizosphere. Some of these PGPR can also enter root interior and establish endophytic populations. Many of them are able to transcend the endodermis barrier, crossing from the root cortex to the vascular system, and subsequently thrive as endophytes in stem, leaves, tubers, and other organs. The extent of endophytic colonization of host plant organs and tissues reflects the ability of bacteria to selectively adapt to these specific ecological niches. Consequently, intimate associations between bacteria and host plants can be formed without harming the plant. Although, it is generally assumed that many bacterial endophyte communities are the product of a colonizing process initiated in the root zone, they may also originate from other source than the rhizosphere, such as the phyllosphere, the anthosphere, or the spermosphere.

Despite their different ecological niches, free-living rhizobacteria and endophytic bacteria use some of the same mechanisms to promote plant growth and control phytopathogens. The widely recognized mechanisms of biocontrol mediated by PGPR are competition for an ecological niche or a substrate, production of inhibitory allelochemicals, and induction of systemic resistance (ISR) in host plants to a broad spectrum of pathogens and/or abiotic stresses.

COMPETITIVE ROOT COLONIZATION

Despite their potential as low-input practical agents of plant protection, application of PGPR has been hampered by inconsistent performance in field tests; this is usually attributed to their poor rhizosphere competence. Rhizosphere competence of biocontrol agents comprises effective root colonization combined with the ability to survive and proliferate along growing plant roots over a considerable time period, in the presence of the indigenous microflora. Given the importance of rhizosphere competence as a prerequisite of effective biological control, understanding root-microbe communication, as affected by genetic and environmental determinants in spatial and temporal contexts, will significantly contribute to improve the efficacy of these biocontrol agents.

Factors Affecting Root Colonization

The root surface and surrounding rhizosphere are significant carbon sinks. Photosynthate allocation to this zone can be as high as 40%. Thus, along root surfaces there are various suitable nutrient-rich niches attracting a great diversity of microorganisms, including phytopathogens. Competition for these nutrients and niches is a fundamental mechanism by which PGPR protects plants from phytopathogens. PGPR reach root surfaces by active motility facilitated by flagella and are guided by chemotactic responses. Known chemical attractants present in root exudates include organic acids, amino acids, and specific sugars. Some exudates can also be effective as antimicrobial agents and thus give ecological niche advantage to organisms that have adequate enzymatic machinery to detoxify them. The quantity and composition of chemoattractants and antimicrobials exuded by plant roots are under genetic and environmental control. This implies that PGPR competence highly depends either on their abilities to take advantage of a specific environment or on their abilities to adapt to changing conditions. As an example, *Azospirillum* chemotaxis is induced by sugars, amino acids, and organic acids, but the degree of chemotactic response to each of these compounds differs among strains. PGPR may be uniquely equipped to sense chemoattractants, e.g., rice exudates induce stronger chemotactic responses of endophytic bacteria than from non-PGPR present in the rice rhizosphere.

Bacterial lipopolysaccharides (LPS), in particular the O-antigen chain, can also contribute to root colonization. However, the importance of LPS in this colonization might be strain dependent since the O-antigenic side chain of *Pseudomonas fluorescens* WCS374 does not contribute to potato root adhesion, whereas the O-antigen chain of *P. fluorescens* PCL1205 is involved in tomato root colonization. Furthermore, the O-antigenic aspect of LPS does not contribute to rhizoplane colonization of tomato by the plant beneficial endophytic bacterium

P. fluorescens WCS417r but, interestingly, this bacterial factor was involved in endophytic colonization of roots.

It has also been recently demonstrated that the high bacterial growth rate and ability to synthesize vitamin B_1 and exude NADH dehydrogenases contribute to plant colonization by PGPR, Another factor of root colonization ability by bacteria is type IV pili, better known for its involvement in the adhesion of animal and human pathogenic bacteria to eukaryotic cells. The type IV pili also play a role in plant colonization by endophytic bacteria such as *Azoarcus* sp.

Root Colonization and Site-specific Recombinase

Bacterial traits required for effective root colonization are subject to phase variation, a regulatory process for DNA rearrangements orchestrated by site-specific recombinase. In certain PGPR, efficient root colonization is linked to their ability to secrete a site-specific recombinase. Transfer of the site-specific recombinase gene from a rhizosphere-competent *P. fluorescens* into a rhizosphere-incompetent *Pseudomonas* strain enhanced its ability to colonize root tips.

Utilization of Root Exudates and Root Mucilage by PGPR

Root exudates are the primary source of nutrients for rhizosphere microorganisms, rhizosphere competence implies that PGPR are well adapted to their utilization. Despite the fact that sugars have often been reported as the major carbon source in exudates, the ability to use specific sugars does not play a major role in tomato root colonization. Similarly, although amino acids are present in root exudates, the bioavailability of amino acids alone is considered insufficient to support root tip colonization by auxotrophic mutants of *P. fluorescens* WCS365. In contrast, Simons et al. reported that amino acid synthesis is required for root colonization by *P. fluorescens* WCS365, indicating that amino acid prototrophy is involved in rhizosphere competence. In addition, PGPR regulate the rate of uptake of polyamines such as putrescine, spermine, and spermidine, since their high titer could retard bacterial growth and reduce their ability to competitively colonize roots. Root mucilage also offers an utilizable carbon source for PGPR to use for the competitive colonization.

ALLELOCHEMICALS FOR BIOCONTROL

Offensive PGPR colonization and defensive retention of rhizosphere niches are enabled by production of bacterial allelochemicals, including iron-chelating siderophores, antibiotics, biocidal volatiles, lytic enzymes, and detoxification enzymes

Siderophores

Iron is an essential growth element for all living organisms. The deficiency of bioavailable iron in soil habitats and on plant surfaces results into furious competition. Under iron-limiting conditions PGPR produce low-molecular-weight compounds called siderophores to competitively acquire ferric ion e.g. *Pseudomonas fluorescens*. Although various bacterial siderophores differ in their abilities to sequester iron, in general, they deprive pathogenic fungi of this essential element since the fungal siderophores have lower affinity. Some PGPR strains go one step further and draw iron from heterologous siderophores produced by cohabiting microorganisms.

Siderophore biosynthesis is generally tightly regulated by iron-sensitive Fur proteins, the global regulators GacS and GacA, the sigma factors RpoS, PvdS, and FpvI, quorum-sensing autoinducers such as *N*-acyl homoserine lactone, and site-specific recombinases. A myriad of environmental factors can also modulate siderophores synthesis, including pH, the level of iron and the form of iron ions, the presence of other trace elements, and an adequate supply of carbon, nitrogen, and phosphorus.

Antibiosis

The antibiosis mechanism of PGPR has become increasingly better understood over the past two decades. A variety of antibiotics have been identified, including compounds such as amphisin, 2,4-diacetylphloroglucinol (DAPG), hydrogen cyanide, oomycin A, phenazine, pyoluteorin, pyrrolnitrin, tensin, tropolone, and cyclic lipopeptides produced by pseudomonads and oligomycin A, kanosamine, zwittermicin A, and xanthobaccin produced by *Bacillus*, *Streptomyces*, and *Stenotrophomonas* spp. Antibiotic synthesis is tightly linked to the overall metabolic status of the cell, which in turn is dictated by nutrient availability and other environmental stimuli, such as major and minor minerals, type of carbon source and supply, pH, temperature, and other parameters. Trace elements particularly zinc, and carbon source levels influence the genetic stability/instability of bacteria, affecting their ability to produce secondary metabolites. It is important to note that many strains produce pallet of secondary antimicrobial metabolites and that conditions favoring one compound may not favor another. Thus, the varied arsenal of biocontrol strains may enable antagonists to perform their ultimate objective of pathogen suppression under the widest range of environmental conditions. For example, in *P. fluorescens* CHAO biosynthesis of DAPG is stimulated and pyoluteorin is repressed in the presence of glucose as a carbon source. As glucose is depleted, however, pyoluteorin becomes the more abundantly antimicrobial compound produced by this strain. This ensures a degree of flexibility for the antagonist when confronted with a different or a changeable environment. Biotic conditions can also influence antibiotic biosynthesis. For example bacterial metabolites salicylates and pyoluteorin can affect DAPG production by *P. fluorescens* CHAO. Furthermore, plant growth and development also influence antiobiotic production, since biological activity of DAPG producers is not induced by the exudates of young plant roots but is induced by the exudates of older plants, which results in selective pressure against other rhizosphere microorganisms. Plant host genotype also plays a significant role in the disease-suppressive interaction of plant with a microbial biocontrol agent.

Lytic Enzyme Production

A variety of microorganisms also exhibit hyperparasitic activity, attacking pathogens by excreting cell wall hydrolases. Chitinase produced by *S. plymuthica* C48 inhibited spore germination and germ-tube elongation in *Botrytis cinerea*. The ability to produce extracellular chitinases is considered crucial for *Serratia marcescens* to act as antagonist against *Sclerotium rolfsii*, and for *Paenibacillus* sp. strain 300 and *Streptomyces* sp. strain 385 to suppress *Fusarium oxysporum* f. sp. *cucumerinum*. It has been also demonstrated that extracellular chitinase and laminarinase synthesized by *Pseudomonas stutzeri* digest and lyse mycelia of *F. solani*. Although, chitinolytic activity appears less essential for PGPR such as *S. plymutica* IC14 when used to suppress *S. sclerotiorum* and *B. cinerea*, synthesis of proteases and other biocontrol traits are involved. The ß-1,3-glucanase synthesized by *Paenibacillus* sp. strain 300 and *Streptomyces* sp. strain 385 lyse fungal cell walls of *F. oxysporum* f. sp. *cucumerinum*.

B. cepacia synthesizes ß-1,3-glucanase that destroys the integrity of *R. solani*, *S. rolfsii*, and *Pythium ultimum* cell walls. Similar to siderophores and antibiotics, regulation of lytic enzyme production (proteases and chitinases in particular) involves the gene regulatory systems and colony phase variation.

Detoxification and Degradation of Virulence Factors

Another mechanism of biological control is the detoxification of pathogen virulence factors. For example, certain biocontrol agents are able to detoxify albicidin toxin produced by *Xanthomonas albilineans*. The detoxification mechanisms include production of a protein that reversibly binds the toxin in both *Klebsiella oxytoca* and *Alcaligenes denitrificans*, as well as an irreversible detoxification of albicidin mediated by an esterase that occurs in *Pantoea dispersa*. Several different microorganisms, including strains of *B. cepacia* and *Ralstonia solanacearum*, can also hydrolyze fusaric acid, a phytotoxin produced by various *Fusarium* species. More often though, pathogen toxins display a broad-spectrum activity and can suppress growth of microbial competitors, or detoxify antiobiotics produced by some biocontrol microorganisms, as a self-defense mechanism against biocontrol agents.

Recently, it has been discovered that certain PGPR quench pathogen quorum-sensing capacity by degrading autoinducer signals, thereby blocking expression of numerous virulence genes. Since most, if not all, bacterial plant pathogens rely upon autoinducer-mediated quorum-sensing to turn on gene cascades for their key virulence factors (e.g., cell-degrading enzymes and phytotoxins), this approach holds tremendous potential for alleviating disease, even after the onset of infection, in a curative manner.

Although biocontrol activity of microorgansims involving synthesis of allelochemicals has been studied extensively with free-living rhizobacteria, similar mechanisms apply to endophytic bacteria, since they can also synthesize metabolites with antagonistic activity toward plant pathogens. Interestingly, the ability to inhibit pathogen growth by endophytic bacteria, isolated from potato tubers, decreases as the bacteria colonize the host plant's interior, suggesting that bacterial adaptation to this habitat occurs within their host and may be tissue type and tissue site specific .

INDUCED SYSTEMIC RESISTANCE IN PLANTS

Biopriming plants with some PGPR can also provide systemic resistance against a broad spectrum of plant pathogens. Diseases of fungal, bacterial, and viral origin and in some instances even damage caused by insects and nematodes can be reduced after application of PGPR.

Induced Systemic Resistance

Certain bacteria trigger a phenomenon known as ISR phenotypically similar to systemic acquired resistance (SAR). SAR develops when plants successfully activate their defense mechanism in response to primary infection by a pathogen, notably when the latter induces a hypersensitive reaction through which it becomes limited in a local necrotic lesion of brown, desiccated tissue. As SAR, ISR is effective against different types of pathogens but differs from SAR in that the inducing PGPR does not cause visible symptoms on the host plant. PGPR-elicited ISR was first observed on carnation (*Dianthus caryophillus*) with reduced susceptibility to wilt caused by *Fusarium* sp. and on cucumber (*Cucumis sativus*) with reduced susceptibility to foliar disease caused by *Colletotrichum orbiculare*. Manifestation of ISR is dependent on the combination of host plant and bacterial strain. Most reports of PGPR-mediated ISR involve free-living rhizobacterial strains, but endophytic bacteria have also been observed to have ISR activity. For example, ISR was triggered by *P. fluorescens* EP1 against red rot caused by *Colletotrichum falcatum* on sugarcane, *Burkholderia phytofirmans* PsJN against *Botrytis cinerea* on grapevine and *Verticllium dahliae* on tomato, *Bacillus pumilus* SE34 against *F. oxysporum* f. sp. *pisi* on pea roots and *F. oxysporum* f. sp. *vasinfectum* on cotton roots.

Factors Affecting ISR

The ability to act as bioprotectants via ISR has been demonstrated for both rhizobacteria and bacterial endophytes, and considerable progress has been made in elucidating the mechanisms of plant-PGPR-pathogen interaction. Several bacterial traits (i.e., flagellation and production of siderophores and lipopolysaccharides) have been proposed to trigger ISR, but there is no compelling evidence for an overall ISR signal produced by bacteria. It has recently been reported that volatile organic compounds may play a key role in this process. For example, volatiles secreted by *B. subtilis* GBO3 and *B. amyloquefaciens* IN937a were able to activate an ISR pathway in *Arabidopsis* seedlings challenged with the soft-rot pathogen *Erwinia carotovora* subsp. *carotovora*. The majority of PGPR that activate ISR appear to do so via a salicylic acid -independent pathway involving jasmonate and ethylene signals. ISR is associated with an increase in sensitivity to these hormones rather than an increase in their production, which might lead to the activation of a partially different set of defense genes.

Defense Mechanisms of ISR-mediated by PGPR

PGPR-triggered ISR fortifies plant cell wall strength and alters host physiology and metabolic responses, leading to an enhanced synthesis of plant defense chemicals upon challenge by pathogens and/or abiotic stress factors. After inoculation of tomato with endophytic *P. fluorescens* WCS417r, a thickening of the outer tangential and outermost part of the radial side of the first layer of cortical cell walls occurred when epidermal or hypodermal cells were colonized. In *Burkholderia phytofirmans* PsJN-grapevine interaction, a host defense reaction coinciding with phenolic compound accumulation and a strengthening of cell walls in the exodermis and in several cortical cell layers was also observed during endophytic colonization of the bacterium. The type of bacterized plant response induced after challenge with a pathogen resulted in the formation of structural barriers, such as thickened cell wall papillae due to the deposition of callose and the accumulation of phenolic compounds at the site of pathogen attack. Biochemical or physiological changes in plants include induced accumulation of pathogenesis-related proteins (PR proteins) such as PR-1, PR-2, chitinases, and some peroxidases. However, certain PGPR do not induce PR proteins but rather increase accumulation of peroxidase, phenylalanine ammonia lyase, phytoalexins, polyphenol oxidase, and/or chalcone synthase. Recent evidence indicates that induction of some of these plant defense compounds (e.g., chalcone synthase) may be triggered by the same *N*-acyl homoserine lactones that bacteria use for intraspecific signaling.

FUTURE PROSPECTS OF PGPR

Researches into the mechanisms of plant growth promotion by PGPR have provided a greater understanding of the multiple facets of disease suppression by these biocontrol agents. Still, most of the focus has been on free-living rhizobacterial strains, especially to *Pseudomonas* and *Bacillus*. Much remains to be learned from nonsymbiotic endophytic bacteria that have unique associations and apparently a more pronounced growth-enhancing effect on host plants.

Revelations about the mechanisms of PGPR action open new doors to design strategies for improving the efficacy of biocontrol agents. Identification of key antimicrobials produced by superior agents, such as 2, 4-diacetylphloroglucinol, can be exploited for streamlining strain discovery by targeting selection of new isolates that carry relevant biosynthetic genes. Determination of the role of edaphic parameters favorable for disease suppression, particularly those that stimulate antibiotic production and activity, can be exploited by targeting inoculants for soils that are more likely to support biocontrol. For example, amending soils or growth substrates with minerals such as zinc or priming inoculants with media amendments during fermentation can be very effective. Similarly, modulation of the rhizosphere bacteria consortia can be accomplished by soil aeration, hydrogenation, and delivery of molasses, sugars and by appropriate crop rotations.

Table 1 : Some examples of PGPR used as Biocontrol agents

Name of PGPR strain	Target Pathogen/Disease
Pseudomonas chlororaphis 63-28	wilt diseases as well as stem and root rots
Bacillus pumilus QST2808	Asian soybean rust
Pseudomonas syringae	*Botrytis cinerea, Penicillium* spp., *Mucor pyroformis, Geotrichum candidu*
Pseudomonas fluorescens A506	Frost damage, *Erwinia amylovora*, and russet-inducing bacteria
*Bacillus subtilis*GB03	*Rhizoctonia, Pythium, Fusarium*, and *Phytophthora*
Burkholderia cepacia	*Rhizoctonia, Pythium, Fusarium*, and disease caused by lesion, spiral, lance, and sting nematodes
Bacillus licheniformis SB3086	Dollar spot, low and moderate disease pressure
Pseudomonas fluorescens strains A506, 1629RS and *Pseudomonas syringae* 742RS	Frost-forming bacteria
Agrobacterium radiobacter Strain 84	Crown gall disease caused by *Agrobacterium tumefacien*
Bacillus pumilus GB34	Fungal pests *Rhizoctonia* and *Fusarium*
Bacillus licheniformis SB3086	Many fungal species especially those causing leafspot and blight diseases

Identifying different mechanisms of action facilitate the combination of strains, bacteria with bacteria or bacteria with fungi, to hit pathogens with a broader spectrum of microbial weapons. Along this same line, biotechnology can be applied to further improve strains that have prized qualities (e.g., formulation ease, stability, or otherwise exceptionally suited to plant colonization) by creating transgenic strains that combine multiple mechanisms of action. For example, transforming the 1-aminocyclopropane-1-carboxylic acid deaminase gene, which directly stimulates plant growth by cleaving the immediate precursor of plant ethylene into *P. fluorescens* CHAO, not only increases plant growth but can also increase biocontrol properties of PGPR. Continued work with endophytic bacteria also holds potential for developing biocontrol agents that may be self-perpetuating by colonizing hosts and being transferred to progeny much as is the case with associative nitrogen-fixing PGPR on sugarcane or the nonsymbiotic endophyte bacterium *Burkholderia phytofirmans* PsJN.

❑❑❑

Chapter **8**

Pest Management in Organic Farming

ARUP DAS*, RAMANUJ VISHWAKARMA* AND MD. WASIM SIDDIQUI**

**Department of Agril. Entomology, Bidhan Chandra Krishi Vishwavidyalaya, Mohanpur, Nadia, W.B.–741252*
***Department of Post Harvest Technology of Horticultural Crops, Bidhan Chandra Krishi Vishwavidyalaya, Mohanpur, Nadia, W.B.–741252*

Before going deep into the pest management strategies in organic farming, let's go through the definition.

"Organic agriculture is a holistic production management system which promotes and enhances agro-ecosystem health, including biodiversity, biological cycles, and soil biological activity. It emphasises the use of management practices in preference to the use of off-farm inputs, taking into account that regional conditions require locally adapted systems. This is accomplished by using, where possible, agronomic, biological, and mechanical methods, as opposed to using synthetic materials, to fulfill any specific function within the system." (FAO/WHO Codex Alimentarius Commission, 1999).

In today's world it's hard to think of managing pest without using the synthetic pesticides. But in organic farming use of synthetic materials is prohibited. So pests should be avoided by good manufacturing practices. This includes general cleanliness and hygiene. Treatments with pest regulating agents must be regarded as the last resort. Physical barriers like sound, ultra-sound, light, and UV-light, traps (incl. pheromone traps and static bait traps), temperature control, controlled atmosphere and diatomaceous earth should be used but before everything a plan for pest prevention and pest control should be developed.

Methods we can take for the purpose of pest management:

- Preventive methods such as disruption, elimination of habitat and access to facilities
- Cultural, mechanical, physical and biological methods

- Pesticidal substances contained in the Appendices of the national standards
- Other substances used in traps
- The certification programme shall set up rules to determine which protection agents and disinfectants may be used.

Some points to remember:

- Irradiation is prohibited.
- There shall never be direct or indirect contact between organic products and prohibited substances. (e.g. pesticides). In case of doubt, it shall be ensured that no residues are present in the organic product.
- Persistent or carcinogenic pesticides and disinfectants are not permitted.

The different methods for pest management are discussed below.

Cultural Control—Cultural practices include all the crop production and management techniques which are utilized by the farmers to maximize their crop productivity and/or farm income. This include practices involved from the very beginning of crop sowing to crop harvesting, and even decisions on crops/varieties to be grown, off-season operations etc. The manipulation of these practices for reducing or avoiding pest damage to crops is known as cultural control. This method is also called *ecological management* or *environmental control*.

Change in sowing parameters—Change in planting time helps to minimize pest damage through asynchronization between the host plant and the pest or to provide the pest its unfavourable situation. The attack of *Helicoverpa armigera* (Hubner) and *Bemisia tabaci* (Gennadius) on cotton can be averted by completion of sowing of cotton within mid-May, as these pests are late season pest. In India, early planting has been found to reduce gall midge and leaf folder damage in rice; shootfly and headbug damage in sorghum and millets; whitegrub damage in groundnut; mustard aphid damage in *Brassica* crops; pod borer damage to chickpea in northern-India. Timely and synchronous planting has been found to reduce bollworm damage in cotton and stem borer damage in sugarcane.

Appropriate seed-rate ensures proper crop stand, spacing and crop canopy that helps in proper intercultural operations. Use of high seed rate is recommended in those crops where removal of infested plants is helpful in minimising the incidence of insect pests, *viz.* Maize borer in maize and sorghum shootfly in sorghum and other crops.

Proper spacing ensures maximum high quality yield per unit area per unit time. But spacing may also influence the population and damage of many insect pests by modifying the micro-environment of the crop or affecting health, vigour and strength of the crop plants or pattern and duration of crop growth and development. Closer spacing has been reported to increase the incidence of planthoppers *viz.* brown planthopper and whitebacked planthopper, gall midge and leaf folder in rice crop. On the other hand, closer spacing resulted in lower incidence of green leafhopper, rice hispa and whorl maggot (Singh and Dhaliwal, 1994). The closer spacing in cotton results in bushy growth of the crop that affects penetration of light, results in vertical growth of the plant, hinders spraying operation and also results in higher relative humidity that favours higher incidence of sucking pests and bollworms. Closer spacing in groundnut lowered the incidence of thrips, jassids, aphids and leafminer, and also increased parasitism in latter. In soyabean, damage by a number of insect pests including *Spilosoma obliqua* Walker, *Melanagromyza sojae* (Zehntner) and

Bemisia tabaci (Gennadius) was greater in closely planted crop. In sugarcane, closer spacing resulted in higher incidence of shoot, internode and stalk borers.

Tillage—The type and timing of tillage can markedly influence the soil environment and affect the survival of insect pests or their natural enemies. Fall ploughing is often helpful in reducing the overwintering population of *Helicoverpa* spp. and several species of cutworms that undergo diapauses in the soil during winter. Deep ploughing, soil raking and hoeing also helps in exposing pests to their natural enemies or unfavourable weather.

Crop rotation—In Central and Southern India intercropping of cotton with black gram, green gram, onion, cowpeas, etc. is reported to divert the population of sucking pests and American bollworm from cotton. Monoculture of cotton was also found to harbour more insect pests than cotton intercropped with groundnut, cowpea and soybean. The intercrop of cowpeas in cotton helped in the colonization of Coccinellids and also enhanced the parasitism of spotted bollworm. On the other hand, okra, mung and pigeonpea as intercrops with cotton increased the population build up of jassid, whitefly, spotted bollworms and American bollworm (Simwat, 1994; Dhawan, 1999).

Physical methods—These methods aim to reduce pest populations by using devices which affect them physically or alter their physical environment.

Hot and cold treatment—Exposure of cotton seeds to sun heat in thin layers for 2-3 days during April helps in killing diapausing larvae of pink bollworm. Treatment of sugarcane setts with heat therapy units either as hot water or hot air treatment kills the scale insects carried over through the setts. Cold storage of fresh and dry fruits and vegetables if often resorted to for escaping fruit fly, potato tuber moth and pathogen damage. A temperature of 10^0C for several days kills fruitfly maggots.

Moisture—Manipulation of prevailing moisture helps in controlling pests, but it is more useful in storages rather than in the field itself. Well-dried grains (moisture content < 8-10%) stored in a cool dry place escape insect damage.

BIORATIONAL AND BIOCONTROL MEASURES

Sex pheromone trap or pheromone trap—Monitoring of pest population is a pre-requisite to efficient pest management strategy. Pheromone trap is one such tool, which is used for monitoring of pests. Sex pheromones are bio-chemicals released usually by female insects as a means of chemical communication to attract males for mating. They are highly specific and are perceived by the males of the same species only. The synthetic pheromones are now in use for luring a variety of lepidopteran pests in small entrapments popularly called pheromone traps. These traps are effectively used in cotton, rice, vegetable and many other crops. Two traps per acre are often sufficient for monitoring the pest population. Pheromones are also exploited for mass trapping of certain pests particularly yellow stem borer in rice. Usually ten traps per acre are used to mass trap the moths to reduce their population.

BIOCONTROL AGENTS

Predators—Predators are free-living organisms and require several preys to complete their life-cycle. In general, they tend to feed on preys smaller than themselves, except when they release a powerful poison, use a trap or hunt in groups. In India, Vidalia beetle was introduced in 1929-'30 to control cottony-cushion scale (origin - Australia) on *Casuarina* and other crops. The beetle established and provided satisfactory control of this pest. Now this insect is established in more than 60 countries, providing satisfactory suppression of cottony-cushion scale. Australian ladybird beetle, *Cryptolaemus montrouzieri* was introduced in India in 1898 for the control of green scale, *Coccus viridis*. It did not establish on green scale but was recorded later on several commercial, horticultural and plantation crops. Coccinellid like *Pharoscymnus horni, Chilocorus nigrita* and *Stichlotis madagassa* have been used on the sugarcane scale insect *Melanaspis glomerata*. The introduction and inoculative release of exotic ladybird beetle, *Curinus coerulens* in 1988 has given spectacular control of exotic psyllid, *Heteropsylla cubana* and saved the *subabul* plantations (primarily used as fodder and fuel) in south India. Common chrysopids like *Chrysoperla carnea, Mallada boninensis* have been used in cotton ecosystem for protection from aphids and other soft-bodied insects. Anthocorid bugs, the minute pirate bugs, play an important role in management of thrips, aphids, eggs and young larvae of moths. Syrphids or hover flies have been recorded as important predators of aphids on several crops.

Parasitoids—A parasitoid is a special kind of parasite which is often about the same size as its host, kills its host and requires only one host (prey) for development into a free-living adult. The adult food of parasitoid is different from that of the larvae. They can be specific or polyphagous. The host specific parasitoids are generally preferred for classical biological control for the management of introduced (exotic) pests. Polyphagous parasitoids are mass produced and used for the management of a variety of lepidopteran tissue-borers and leaf feeders. Trichogrammmatid egg parasitoids have been used during the active egg-laying period for management of *Chilo* spp., *Scirpophaga excerptalis* and *Acigona steniellus* on sugarcane, *C. partellus on* maize, *S. incertulus* on rice, *Helicoverpa armigera, Earias* spp. and *Pectinophora gossypiella* on cotton and *H. armigera, Agrotis ipsilon, A. segetum* and *A. biconica* on several crops. They are released @ 50,000/ha on sugarcane, @ 75,000/ha on maize, @ 50,000-1,00,000/ha on rice and @ 1,50,000/ha/release on cotton. The thumb rule is that the dosage should be adjusted in such a way to provide 4-6 *Trichogramma*m^2. *Trichogramma chilonis* has been released successfully in Punjab, Haryana, Gujarat, Uttaranchal, Uttar Pradesh, Bihar, Karnataka, Tamil Nadu and Maharashtra. It has given 56-82% protection against the sugarcane tissue-borers. The *T. chilonis* has been released area-wise on sugarcane, cotton, sunflower, sorghum, groundnut and maize against *Chilo* spp. and *Scirpophaga incertulas*, and *T. chilonis* BioH3/*T. pretiosum*/*T. bresiliensia* against *H. armigera* on several crops. *Epiricania melanoleuca* has been used for the control of *Pyrilla perpusilla* on sugarcane.

Entomopathogens—Entomopathogens are disease-causing organisms (such as bacteria, viruses, fungi and protozoa in insect pests), which either kill their hosts or debilitate the future generations. Under certain situations, they cause disease epizootics in the field, e.g. *Verticillium lecanii* on scale insects, nucleopolyhedrovirus or *Nomuraea rileyii* in lepidopterans. The infected insects are unable to feed properly, remain stunted, lose their body colour, and get immobile or paralyzed. Dead insect pests are often found hanging on the plants, or a fungal growth on the body of the dried insect could be easily located in the field.

Entomopathogenic bacteria—They are extra-cellularly associated with insects and enter the body cavity through contaminated feed and multiply. Bacteria also invade through integument of the insect injured by pricking of parasitoids and infections caused by protozoa or nematodes. Death of the insect may either occur due to intoxication, sudden lack of oxygen, chemical changes in the gut or by the toxins or crystals produced in the bacterial cells. In India, *Bacillus thuringiensis* subsp. *kurstaki* was isolated from *Helicoverpa armigera* from Mysore, Karnataka in 1955. Subsequently *B. thuringiensis* subsp. *kenyae* (ISPC-1, ISPC-7), *Bacillus thuringiensis* subsp. *tolworthi* (HD-2), *Bacillus thuringiensis* subsp. *thuringiensis* (HD-263), *Bacillus thuringiensis* subsp. *kurstaki* (HD-709, ISPC-4, H3a, 3b, 3c) *Bacillus thuringiensis* subsp. *aizawai* (HD-549, ISPC-1), *Bacillus thuringiensis* subsp. *tenebrionis* and *Bacillus thuringiensis* subsp. *israelensis* (VCRC B-17, B-113, IPS-80, H-14) have been identified. Many *Bacillus thuringiensis* subsp. infect lepidopterans, but *Bacillus thuringiensis* subsp. *tenebrionis* infects coleopteran and *Bacillus thuringiensis* subsp. *israelensis* (VCRC B-17, B-113, IPS-80, H-14) infect dipteran insects. *Bacillus thuringiensis* and its subspecies have been recorded infecting more than 68 species of insects, mainly lepidopterans. *Paenibacillus papillae* is produced by infecting live white-grubs. It has been registered in USA since about 1950, the first registration of any insect pathogen as a microbial control agent.

Baculoviruses—Most insect viruses end up with proteinic occlusion bodies (POBs) of different sizes and shapes in different organs or cells of the infected host insect. Currently, insect pathogenic viruses such as nucleopolyhedroviruses (NPVs) and granuloviruses (GVs) are widely used in pest control. They are generally very specific to host species or genus. The NPVs are irregular, spherical, cubic or hexagonal crystals with blunt or melted edges. Each polyhedron measures 3-10 microns in diameter and contains many particles or virus rods. For the control of *H. armigera* on chickpea, groundnut, Egyptian clover, tobacco and sunflower, *Ha*-NPV is recommended at 1.5-3.0 X 10^{12} POB/ha (250 LE/ha); for pigeonpea 1.5-3.0 X 10^{12} POB/ha (250-500 LE/ha) and for cotton 3.0 X 10^{12} POB/ha (500 LE/ha) is recommended. Based on *H. armigera* population, 2-4 sprays are given. The virus suspension can also be safely and effectively combined with several insecticides at reduced dosages like neem seed-kernel extract (NSKE) for increasing the efficacy. It has also been used in combination with egg parasitoid, *Trichograma* spp. and predator *Chrysoperla carnea*. For the control of *Spodoptera litura* on crops such as tobacco, groundnut, blackgram and cotton, *Sl*-NPV is given as spray as soon as the eggs hatch and subsequent sprays are carried out at 7-10 days interval. The dosage varies from 1.5 X 10^{12} to 3.0 X 10^{12} POB/ha (250 LE to 500 LE/ha). The GV infected insects contain 0.3 to 0.5 micron long oval granules. In India over 16 GVs have been recorded from their respective insect hosts. GV of sugarcane shoot-borer *Chilo infuscatellus* has been successfully used against this pest on sugarcane.

Entomofungi—The fungi attacking insects belong to class Phycomycetes, Ascomycetes, Basidiomycetes and Deuteromycetes. Only about 20 species have received more attention for use against insect pests, as these are amenable for biological control, being easy to produce and relatively low priced. Entomofungal pathogens are popularly known by different names, depending on the colour of the fungus as white muscardine, green muscardine, brown muscardine, red muscardine, yellow muscardine or even black muscardine. Among Entomofungal pathogens, *Beauveria bassiana* against rice hispa, *Dicladispa armigera* and lepidopteran tissue-borers of sugarcane, *Metarhizium anisopliae* and *Beauveria brongniartii* against *Holotrichia serrata* and *Leucopholis lepidophora* on sugarcane and *Holotrichia consanguinea* on groundnut, pigeonpea, cotton etc. and *Helicoverpa armigera* on cotton, groundnut, pigeonpea have been found promising. Natural epizootics of *N. rileyi* were recorded from *H. armigera* and *S. litura* from cotton, groundnut, pigeonpea and other crops.

Entomopathogenic nematodes—Entomopathogenic nematodes (EPNs) (steinernematids and heterorhabditids) directly attack the insect pests. The invasive larve of EPNs carry the associated bacteria *Xenorhabdus* sp.or *Photorhabdus* sp. in their gut, which not only kill the invaded host within 24-48 hrs but also produce a biosatic substance that retards the putrification of the cadaver. The nematodes feed on these bacteria and are capable of producing several generations in their dead hosts. The EPNs are susceptible to desiccation, but are important natural mortality factors of cutworms (*Agrotis* spp.) in irrigated fields. The EPNs are promising against several insect pests in cryptic habitats such as whitegrubs, gram pod-borer, rice stem-borer and other lepidopterans. Since 1966 extensive countrywide explorations have been conducted to quantify the diversity of EPNs in the country. *Steinernema bicornutum* (PDBC EN 3.2), *S. carpocapsae* (PDBC EN 16.61, 6.11, 1.3, 1.4, 7.2), *S. abbasi* (PDBC EN 3.1), *S. tami* (PDBC EN 2.1), *Steinernema* sp. (SSL2) (PDBC EN 13.21, 14.1), *Steinernema* sp. (PDBC EN 11.1), *S. thermophilum, S. feltiae, Heterorhabditis bacteriophora, H. indica* (PDBC EN 7.1, 13.3, 13.4, 13.22, 14.2, 6.71, 14.3) and others have been recorded. *S. carpocapsae, H. indica* and *H. bacteriophora* are promising against white-grubs.

Protozoa—Several classes of protozoa parasitize insects but only Sporozoa and Cnidospora cause death of the parasitized hosts. The infection takes place within 72 hrs of ingestion of contaminated food and water. Braconid and ichneumonid parasitoids transmit the spores of microsporidia through their contaminated ovipositor. Transovarial transmission also occurs. *Mettesia* sp. and *Vairimorpha* sp. are potential parasites of lepidopterans such as *Helicoverpa armigera.* Schizogregarine, *Fairnocystis tribolii* has potential for suppression of stored grain pest, red flour beetle. A species of schizogregarine has potential for suppression of white-grubs.

Rickettsiae—Rickettsiae infect the fat body of insects such as flies (Diptera), beetles (Coleoptera), and grasshoppers (Orthoptera) and destroy the tissues. In the field, the Rickettsiae-infected insects show white staining of their body and the haemolymph turns milky white.

Table 1 : Biocontrol agents that can be applied in the field

Biocontrol agents	Target pest	Crops
Egg parasitoids		
Trichogramma chilonis	*Chilo partellus* *Sesamia inferens*	Sorghum and maize
	Chilo sacchariphagus indicus *Chilo auricillus*	Sugarcane
	Helicoverpa armigera	Tomato, cotton, vegetables & pulses
Trichogramma exiguum	*Chilp sp.* *Helicoverpa armigera*	Sorghum, maize and sugarcane Cotton and vegetables
Trichogramma brasiliensis	*Helicoverpa armigera* *Pectinophora gossypiella*	Cotton and tomato

contd. ...

... contd.

Biocontrol agents	Target pest	Crops
Trichogramma japonicum	*Tryporyza incertulus*	Rice
	Tryporyza novella	Sugarcane
Trichogrammatoidea bactrae	*Plutella xylostella*	Cabbage, cauliflower
Telenomus remus	*Spodoptera litura*	Vegetable and cotton
	Spodoptera exigua	Vegetable and chickpea
Egg-larval parasitoids		
Chelonus blackburni	*Earias sp.*	Cotton and okra
	Pectinophora gossypiella	
	Phthorimoea operculella	Potato
Capidosoma koehleri	*Phthorimoea operculella*	Potato
Larval parasitoids		
Apanteles angaleti	*Pectinophora gossypiella*	Cotton
Cotesia plutellae	*Plutella xylostella*	Cabbage, cauliflower
Predators		
Cryptolaemus montrouzieri	*Maconellicoccus hirsutus*	Guava, sapota, crotons and other ornamentals
Scymnus coccivora	*Ferrisia virgata*	Guava, sapota, crotons and other ornamentals
Nephus regularis	*Planococcus sp.*	Guava, sapota, crotons and other ornamentals
Chrysoperla carnea	Soft bodied insects such as aphids	Cotton
Chilomenes sexmaculata	*Aphis craccivora, Aphis gossypii*	Bean and cotton
Brumoides suturalis	*Mealy bugs* *Aphis craccivora* *Aphis sp.*	Fruits and cotton
Pathogens		
Bacillus thuringiensis	Lepidopterous larvae	All crops
Ha NPV	*Helicoverpa armigera*	Cotton, chickpea and pigeonpea
Sl NPV	*Spodoptera litura*	Vegetable and cotton

Pest Management Strategies for Some Very Common Crops

1. Pests of Rice

Green leafhopper

Management

- Use resistant varieties like IR 50, CR 1009, Co 46, PTB 2 and PTB 18.
- Nursery should not be raised near the lamp posts
- Apply neem cake @ 12.5 kg/20 cent nursery as basal dose
- Maintain the water level at 2.5 cm for 3 days

Brown plant leafhopper (*Nilaparvata lugens)*

Management

- Use resistant varieties like Aruna, Karnataka, Karthika, Krishnaveni, Makon,
- Abhey, Asha, Divya, Py 3, Co 42, Ptb 33 and Ptb 21.
- Avoid close planting and provide 30 cm rogue spacing at every 2.5 m to reduce the pest incidence.
- Control irrigation by intermittent draining
- Set up light traps to monitor pest population and to control
- Release of natural enemies like *Lycosa pseudoannulata, Cyrtorhinus lividipennis*
- Drain the water before the use of insecticides and direct the spray towards the base of the plants
- Spray neem seed kernel extract 5% (25 kg/ha) (or) neem oil 2% (10 l/ha)

White backed planthopper (*Sogatella furcifera)*

Management

- Release of egg parasitoid, *Anagrus* sp. and adults and nymphs of the dryinid
- *Pachygonatopus* sp. Predators recorded on the insect are *Coccinella arcuata, Cyrtorrhinus lividipennis, Tytthus parviceps*

Mealybug (*Brevennia rehi)*

Management

- The mealy bug is parasitised by *Adelencyrtus* sp., *Dolihoceros* sp., *Gyranusa* sp., *Parasyrphophagus* sp., *Xanthoencyrtus* sp and predated upon by *Gitonides perspicax, Leucopis luteicomis, Scymnus* sp., *Pullus* sp., *Anatrichus pygmaeus, Mepachymerus ensifer*.
- Remove the grasses from the bunds and trim the bunds during the main field preparation before transplanting.
- Remove and destroy the affected plants.

Rice black bug (*Scotinophora lurida* and *S. coarctata*)

Management

- Keep the field free from weeds and grasses
- Drain the excess water from the field
- Set up light trap during the full moon period to attract large number of bugs and kill
- Conserve the predators *viz.*, spiders, coccinellids and wasps to check the pest
- Ducks can be allowed in the field to pick up the bugs

Rice earhead bug (*Leptocorisa acuta*)

Botanical powder formulations

- Neem seed kernel extract 5%
- Notchi leaf powder extract 5%
- Ipomoea leaf powder extract 5%
- Prosopis leaf powder extract 5%

Paddy stem borer (*Scirpophaga incertulas*)

Management

- Grow resistant varieties like Ratna, Jaya, TKM 6, IR 20 and IR 26
- Clip the tip of seedlings before transplanting to eliminate egg masses
- Avoid close planting and continuous water stagnation at early stages
- Collect and destroy the egg masses
- Pull out and destroy the affected tillers
- Set up light traps to attract and kill the moths
- Harvest the crop upto the ground level and disturb the stubbles with plough immediately after the harvest
- Release the egg parasitoid, *Trichogramma japonicum* on 30 and 37 DAT twice @ 5 ml/ha/release followed by monocrotophos 36 SL spray thrice @ 1000 ml/ha on 58, 65 and 72 DAT reduce the damage.
- Apply *Bacillus thuringiensis* var *kurstaki* and neem seed kernel extract in the combination of 2.5 kg/ml and 1% to reduce the oviposition by the stemborer.

Gall midge (*Orseolia oryzae*)

Management

- Encourage early planting of the crop to escape from infestation
- Use resistant varieties like MDU-3, Shakthi, Vikram, Sureka
- Harvest the crop and plough immediately
- Remove the alternate host and adjust the time of planting
- Use quick growing varieties
- Set up light trap @ 1 / ha as a monitoring device

- Infra red light trap attracts gall midge effectively
- Release larval parasitoid: *Platygaster oryzae* through parasitized galls @ 1 per 10 m2 in the main field at 10 DAT.
- The carabid beetle, *Ophionia indica* is an effective predator.
- Conserve the spider predators like *Tetragnatha* and *Argiope catenulata* in rice ecosystem to feed on the adult midge.

Swarming caterpillar (*Spodoptera mauritia*)

Management

- Release larval parasitoids *Apanteles ruficrus, Meteorus* sp., *Charops bicolor,*
- *C. dominans, Drino unisetosa, Pseudoperichaeta orientalis, Strobliomyia aegyptia, Pseudogonia cinerascens, Tachinia fallan, Cuphocera varia, Sturmia inconspicua, Chelonus* sp., *Euplectrus euplexiae, E. spodopterae.*
- Pupa by *Netelia* sp., *Actia* sp., *Drino* sp. and *Isomera cinerascens, Hexamermis* sp. a parasitic nematode parasitises the larva.
- The vertebrate predators of the larvae are *Corvus splendens, C. macrorhynchus, Bubalcus coromandus, Ardeela grayi, Amaurovius phoenicocurus, Acridotheres tristis.*
- Flood the nursery to expose the hiding larvae to the surface and thus birds pick
- them up.
- Kerosenate the water while irrigation to suffocate and kill the larvae.
- Allow the ducks into the field to feed on the larvae.

Rice case worm (*Nymphula depunctalis*)

Management

- Release larval parasitoids *viz., Elasmus* sp., *Apanteles* sp., *Bracon* sp., *Hormiues* sp.
- Release pupal parasitoids *viz., Pediobius* sp., *Apsilops* sp., *Eupteromalus parnarae.*
- Drain water from the field.
- Dislodge the cases by running a rope over the young crop.

Leaf folder (or) leaf roller (*Cnaphalocrocis medinalis*)

Management

- Use resistant varieties like TNAU LFR 831311, Cauveri, Akashi, TKM-6, IET 7511, IET 9225 and IET 9797.
- Clipping of affected leaves reduces the pest population.
- Keep the bunds clean by trimming them and remove the grassy weeds.
- Avoid use of excessive nitrogenous fertilizer.
- Set up light traps to attack and kill the moths.
- Release *Trichogramma chilonis* on 37, 44 and 51 DAT thrice followed by three sprays of monocrotophos 36 SL @ 1000 ml/ha on 58, 65 and 72 DAT.
- Spray neem seed kernel extract 5% @ 25 kg.

Grasshopper (*Hieroglyphus banian*)

Management

- Expose the eggs to be picked up by birds after ploughing and trim the bunds
- Bio-control agents such as *Cacallus* spp., *Barycomus* spp. and *Seelio* spp., which are egg parasites should be encouraged.

Short horned grasshopper (*Oxya nitidula*)

Management

- Expose the eggs during summer ploughing and they will be picked up by birds.

Spiny beetle / Rice hispa (*Dicladispa armigera*)

Management

- The leaf tips containing blotch mines should be plucked and destroyed.
- Manual collection and killing of beetles with hand nets may help in reducing the population of the pest.

Integrated Pest Management in Rice

A. Cultural method

- Remove / destroy stubbles after harvest and keep the field free from weeds.
- Trim and plaster the bunds of rice field to expose the eggs of grasshoppers and to eliminate the bugs breeding in grasses.
- Form the buds narrow and short to reduce the damage by rodents.
- Use resistant varieties wherever available.
- Provide effective drainage wherever there is problem of BPH.
- Clip the tip of seedling before transplanting to prevent the carryover of egg masses of rice yellow stem borer from nursery to main field.
- Organize synchronized planting wherever possible.
- Leave 30 cm rogue space at every 2.5 m to reduce damage by BPH and rodents.
- Avoid use of excessive nitrogenous fertilizers.
- Use irrigation water judiciously (Alternative webbing x drying reduce BPH and case worm).
- Remove the egg masses of stem borer in the main field.

B. Mechanical methods

- Dig out the rat burrows and destroy the rats and young ones at the beginning of the season.
- Set up light traps to monitor and control pests.
- Set up-bow traps to kill rodents.

C. Biological methods

- Release *Trichogramma japonicum* on 30 and 37 DAT twice 5 ml/ha/release against stem borer.

- Release *Trichogramma chilonis* on 37, 44 and 51 DAT (Thrice) @ 5 ml/ha/release against leaf folder.
- Release of *Platygaster oryzae* parasitized galls @ 1 per 10 m2 in the main field on, 10 DAT against gall midge.
- Set up owl perches to reduce rat damage.

D. Plant products

- Spray neem seed kernel extract 5% (25 kg/ha), neem oil 3% (15 lit/ha) to control brown planthopper.
- Spray botanical powder formulation *viz.*, NSKE, *Vitex negundo* (Notchi), *Prosopis juliflora* and *Ipomoea carnea* leaf extract 5% to control earhead bug and black bug.

2. Pests of Cotton

Leafhopper (*Amrasca devastans*)

Management

- Use resistant varieties like MCU 3, MCU 5 and MCU 9.
- Setup light trap to monitor the broods of leaf hopper to attract and kill.
- Release predators viz *Chrysopa carnea.*
- Early sowing and close spacing of cotton reduces pest infestation particularly, if the rainfall is heavy.

Cotton aphid (*Aphis gossypii*)

Management

- Release biocontrol agents *viz.*, Coccinellid beetle *Monochilus sexmaculatus* and *Coccinella septumpunctata* and *Aphelinus mali, A. flavipes.*
- **Predators** - *Phylloscopus tristis*
- Monitor the nymphs and adults of early season sucking pests from the 14th day after sowing.

Thrips (*Thrips tabaci*)

Management

- Monitor the nymphs and adults of early season sucking pests from the 14th day after sowing.

Whitefly (*Bemisia tabaci*)

Management

- Use g with recommended spacing, preferably wider spacing is essential and late white fly tolerant varieties like LPS 141 and Supriya.
- Timely so win sowing may be avoided.
- Avoid the alternative, cultivated host crops of the white fly in the vicinity of the cotton crop. Cultivation of brinjal, bhendi, tomato and tobacco near the cotton field may be avoided.

- Grow cotton only once in a year either in winter or summer season in any cotton tract.
- Adopt crop rotation with non-preferred host such as sorghum, ragi, maize etc., to check the buildup of the pest.
- Remove and destroy alternate weed hosts like *Abutilon indicum, Solanum nigrum* from the fields and neighbouring areas.
- Field sanitation may be given proper attention.
- Judicious irrigation management and nitrogenous fertilizer application may be followed to arrest the excessive vegetative growth and pests build up.
- Monitor the activities of the adult white flies by setting up yellow pan traps and sticky traps at 1 foot height and also *in situ* counts.
- Collect and remove white fly infested leaves from the plants and those which were shed due to the attack of the pest and destroying them.
- Spray NSKE 5% and neem oil 5 ml

Red cotton bug (*Dysdercus cingulatus*)

Management

- Biocontrol agent *Harpactor costalis* predaceous on nymph and adult
- Since the pest feeds gregariously, the infested leaves or bolls can be shaken in water and drowned.
- Plough the field to expose the eggs.

Stem weevil (*Pempherules affinis*)

Management

- The infested plants should be removed and destroyed.
- Use resistant variety MCU 3 against stem weevil.
- Apply farm yard manure at 25 t or 250 kg of neem cake/ha.

Tobacco Cutworm (*Spodoptera litura*)

Management

- Use of light trap to monitor and kill the attracted adult moths. Set up the sex pheromone trap Pherodin S.L. at 12/ha to monitor the activity of the pest and to synchronise the pesticide application, if need be, at the maximum activity stage.
- Growing castor along border and irrigation bunds. Removal and destruction of egg masses in castor and cotton crops.
- Removal and destruction of early stage larvae found in clusters which can be located easily from a distance.
- Collection and destruction of shed materials.
- Hand picking and destruction of grown up caterpillars.

Green bollworm (*Helicoverpa armigera*)

Management

- Premonsoon sowing during 4th week of September significantly lower the bollworm damage in rainfed areas of Virudhunagar and Thoothukudi districts.

Other management practices

- Pest monitoring through light traps, pheromone traps and in situ assessments by roving and fixed plot surveys has to be intensified at farm level, village level, and block level, regional and state levels. For bollworm, *H. armigera* management, an action threshold of one egg per plant or 1 larva/ plant may be adopted.

Cultural practices

- Synchronized sowing of cotton preferably with short duration varieties in each cotton ecosystem.
- Avoiding continuous cropping of cotton both during winter and summer seasons in the same area as well as ratooning.
- Avoiding monocropping. Growing of less preferred crops like green gram, blackgram, soyabean, castor, sorghum etc., along with the cotton as intercrop or border crop or alternate crop to reduce the pest infestation.
- Removal and destruction of crop residues to avoid carry over of the pest to the next season, and avoiding extended period of crop growth by continuous irrigation.
- Optimise the use of nitrogenous fertilizers which will not favour the multiplication of the pest.
- Judicious water management for the crop to prevent excessive vegetative growth and larval harborage.

Biological control

- Application of nuclear polyhedrosis virus (NPV) at 3 x 10 12 POB /ha in evening hours at 7th and 12th week after sowing.
- Conservation and augmentation of natural predators and parasites for effective control of the pest.
- Inundative release of egg parasite, *Trichogramma* spp., at 6.25 cc/ha at 15 days interval 3 times from 45 DAS egg-larval parasite, *Chelonus blacksburnii* and the predator*Chrysoperla* 1,00,000/ha at 6th, 13th and 14th week after sowing.

3. Pests of Sugarcane

Shoot borer (*Chilo infuscatellus*)

Management

- Resistant varieties: CO 312, CO 421, CO 661, CO 917 and CO 853
- Planting – December – January escapes the incidence
- Intercrop: Daincha – low shoot borer incidence

- Earthing up – 30th day
- Trash mulching: 10 – 15 cm thickness on 3 days after planting
- Ensure adequate moisture
- Remove and destroy dead hearts
- Tachinid parasite: *Sturmiopsis inferens* @ 125 gravid females

Stem or internode borer

Management

- Resistant varieties: CO 975, CO 7304 and COJ 46
- Collect and destroy the eggs periodically
- Detrash: 150 and 210 DAP
- Avoid use of excessive nitrogen fertilizers
- Egg parasitoid: *Trichogramma chilonis* @ 2.5 m.l / ha – 6 releases – 4th month onwards at 15 days intervals
- Larval parasitoids: *Stenobracon deesae, Xanthopimpla nursei, Apanteles flavipes*
- Pupal parasitoids: *Tetrastichus ayyari, Trichopilus diatraeae, Xanthopimpla stemmato*

Top shoot borer

- Dead heart in grown up canes; dead heart reddish brown in colour
- Parallel rows of shot holes in the emerging leaves and red tunnels in the midribs of leave
- Resistant varieties: CO 419, CO 745 and CO 6516
- Tolerant varieties: CO 859, CO1158 and CO 7224
- Collect and destroy the egg masses
- Ichneumonid parasitoid:*Isotima javensis* @ 100 pairs/ ha (prepupal parasitoid)
- Egg parasitoids: *Telenomus beneficiens, Tetrastichus schoenobi, Trichogramma minutum*
- Larval parasitoids: *Goniozus indicus, Chelonus* sp.
- Pupal parasitoid: *Tetrastichus ayyari*

Leafhopper

Management

- Detrash on the 150th and 210 th day of planting.
- Release lepidopteran parasite *Epiricania melanoleuca* @4-5 lacs egg/cocoon/h

Scale insect

Management

- Detrash as per schedule
- Drain excess water
- Release lady bird beetles *Chilocoris nigritus* or *Phyarascymnus horni*

Biological Control

Shoot borer can be controlled by releasing 76 gravid females of *Sturmiopsis inferens*/ ha. Release egg parasitoids, *Trichogramma australicum* or *T.chilonis* 6 times @ 1cc/release (50,70, 80,90,100,110 days after planting) to control internode borer. Repeat the release from 50th day at fortnightly for the control of internode borer. On 30th and 45th day after planting, apply granulosis virus at 108 IBS ml (300 diseased larvae/ac) in 200 lit/acre or release 50 gravid females of *Sturmiopsis inferns*. Top borer can be successfully controlled by releasing *Isotima javensis* at 40 pairs/ac. Release of *Epiricania melanoleuca* egg masses and cocoon 4-lacs/ha on for managing Pyrilla.

4. Pests of Green Gram

Gram caterpillar (*Helicoverpa armigera*)

Management

- Pheromone traps for *Helicoverpa armigera* 12/ha
- Bird perches 50/ha
- Hand picking of larvae
- Ha NPV 1.5 x1012 POB/ha with teepol (1 ml/lit.)
- Apply any one of the following (Spray fluid 625 l/ha)
- Neem seed kernel extract - 5% twice followed by Neem oil - 12.5 l/ha

(Note: Insecticide / Ha NPV spray should be applied when the larvae are in early stage)

Storage pests (*Callosobruchus chinensi, C. maculates*)

Management

- Dry the seeds adequately to reduce moisture level to 10 %.
- Use pitfall traps or two in one model trap to assess the time of emergence of field carried over pulse beetle in storage and accordingly sun-dry the produce.
- Seed: Mix any one of the following for every 100 kg : TNAU Neem oil 60 EC (C) - 1 lit., Pungam oil - 1 lit.
- Pack in polythene lined gunny bags for storage.

5. Pests of Red gram

Pod borers:Spotted pod borer (*Maruca testulalis*); Grampod borer (*Helicoverpa armigera*); Pod fly (*Melanagromyza obtuse*); Pod bug (*Clavigralla gibbesa*)

Management

- Pheromone traps for *Helicoverpa armigera* 12/ha
- Bird perches 50/ha
- Mechanical collection of grown up larva and blister beetle
- Ha NPV 3 x1012 POB/ha in 0.1% teepol
- NSKE 5% twice followed by triazophos 0.05%
- Neem oil 2%

Note: Insecticide / Ha NPV spray should be made when the larvae are up to third instars

6. Pests of Black gram

Tobacco cut worm (*Spodoptera litura)*

Management

- Use of light trap to monitor and kill the attracted adult moths. Set up the sex pheromone traps at 12/ha to monitor the activity of the pest and to synchronise the pesticide application, if need be, at the maximum activity stage.
- Growing castor along borders.
- Removal and destruction of egg masses in castor and cotton crops.
- Removal and destruction of early stage larvae found in clusters which can be located easily even from a distance.
- Hand picking and destruction of grown up caterpillars.
- Spraying NPV at 1.5 x 1012 POB/ha with teepol (1ml/ lit.)
- Use of poison bait pellets prepared with rice bran 12.5 kg, jaggery 1.25 kg, carbaryl 50 WP 1.25 kg and water 7.5 litres. This bait can be spread in the fields in the evening hours so that the caterpillars coming out of the soil, feed and get killed.

Storage pests (*Callosobruchus chinensi, C. maculates)*

Management

- Dry the seeds adequately to reduce moisture level to 10 %.
- Use pitfall traps or two in one model trap to assess the time of emergence of field carried over pulse beetle in storage and accordingly sun-dry the produce.
- Seed: Mix any one of the following for every 100 kg : TNAU Neem oil 60 EC (C) - 1 lit Pungam oil - 1 lit.
- Pack in polythene lined gunny bags for storage

7. Pests of Pearl Millet

Stem borer (*Chilo partellus*)

Management

- Set up light trap till midnight to attract and kill the stem borer moth.
- Collect the stubbles after harvest and burn to destroy diapausing borers.

8. Pests of Sorghum

Shootfly (*Atherigona varia soccata)*

Management

- Use resistant varieties like Co-1, CSH 15R, Maldandi and Hagari.
- Sow the sorghum immediately after the receipt of monsoon rain to minimise the shoot fly damage.
- Use higher seed rate (12.5 kg/ha) and remove the shoot fly damaged seedlings at the time of thinning or raise nursery and transplant only healthy seedlings.

- Pull out plants showing dead hearts and destroy at the time of thinning.
- Set up hanging type of plastic fishmeal trap @ 12/ha till that is 30 days old.
- Release larval parasitoids *viz., Tetrastichus nyemitavus, T. coimbaborensis, Callitula* sp., *Diaulinopsis* sp.,
- Pupal parasitoid - *Cratalpiella* sp.
- Predator - *Abrolophus* sp.

Stem borer (*Chilo partellus*

Management

- Dead hearts should be pulled out and used as fodder (or) burried in manure pits.
- The stubbles should be ploughed up during winter and burnt to destroy the hibernating larvae.
- Resistant varieties against jowar stem borer should be encouraged.
- Sow the lablab or cowpea as an intercrop to minimise the stem borer damage (Sorghum: Lablab 4:1)
- Set up light trap till midnight to attract and kill the stemborer moths.
- Bio-control agents *viz.,Trichogramma minutum, Bracon chinensis* (egg parasites) and *Microbracon chilocida, Apanteles flavipes, Xanthopimpla* spp. (larval parasites should be encouraged).

Pink stem borer (*Sesamia inferens*)

Management

- Release egg parasitoids: *Telenomus* sp., *Trichogramma minutum*
- Larval parasitoids: *Apanteles flavipes, Bracon drinensis*
- Pupal parasitoids: *Tetrastichus ayyari*

Grain midge (*Contarinia sorgicola*)

Management

- Sett up of light traps till mid night to monitor, attract and kill adults of stem borer, grain midge and earhead caterpillars.

Shoot bug (*Peregrinus maidis*)

Management

- Release egg parasitoids *viz., Paranagrus optabilis, Octetrastichus indicus*
- Predators - *Coccinella septumpunctata, Menochilus sexmaculatus, Geocoris tricolor*

Mirid bug (*Creontiades pallidifer*)

Management

- Release larval parasitoids *Apanteles* sp., *Eupelones popa*
- Larval and pupal parasitoid *Tetrastichus* spp.
- Predators - *Orius albidipennis; Tapinoma indicum*

Earhead caterpillar (*Helicoverpa armigera)*

Management

- Set up of light traps till mid night to monitor, attract and kill adults of stem borer, grain midge and ear head caterpillars.
- Set up sex pheromone traps at 12/ha to attract male moths of *Helicoverpa armigera* from flowering to grain hardening. Two applications of NPV at 10 days interval at 1.5 X1012 POB along with crude sugar 2.5 kg + cotton seed kernel powder 250 g on the earheads is effective in reducing the larval population of *Helicoverpa*.

Pests of Tomato

Helicoverpa armigera

- Collect and destroy the infected fruits and grown up larvae.
- Grow less susceptible genotypes Rupali, Roma, Pusa red plum.
- Grow simultaneously 40 days old American tall marigold and 25 days old tomato
- seedling at 1:10 rows to attract *Helicoverpa* adults for egg laying.
- Setup pheromone trap with Helilure at 15/ha and change the lure once in 15 days.
- Spray HaNPV at 500 LE/ha along with cotton seed oil 300 g/ha to kill larvae.
- Release *Chrysoperla carnea* at weekly interval at 50,000 eggs or grubs / ha from 30
- days after planting.
- Six releases of *T. chilonis* @ 50,000/ha per week coinciding with flowering time and
- based on ETL.
- Spray *B. thuringiensis* 2 g/lit
- Do not spray insecticides after maturity of fruits.
- Encourage activity of parasitoid *Eucelatoria bryani*, *Campoletes*, *Chelonus* etc.,

2. Leaf miner

- Spray NSKE 3%
- Collect and destroy mined leaves

3. *Spodoptera litura*

- Plough the soil to expose and kill the pupae.
- Grow astor along border, irrigation channel as indicator or trap crop.
- Flood the field to drive out hibernating larvae.
- Set up light trap at 1/ha or pheromone trap at 15/ha with pheroclin SL lure.
- Destroy egg masses and grown up larvae.
- Spray SNPV at 250 LE along with teepol 1ml/lit in evening hrs.

4. Whitefly

- Uproot and destroy the diseased leaf curl plants
- Remove alternate weed hosts.
- Use irrigation judiciously.
- Use yellow sticky trap to attract and bill insects.
- Spray FORS 2% / neem oil 0.5% along with teepal 1 ml/lit.
- Encourage activity of parasitoids *Eretmocerus mesii* and predator coccinellids,
- Brumus and Chrysoperla.

5. Thrips

- Collect and destroy the damaged leaves and twigs and uproot the diseased plants.
- Use yellow sticky traps at 15/ha to attract and kill insects.
- Release first instar larvae of *Chrysoperla carnea* @ 10,000/ha and encourage
- coccinellid predator.

SUBSTANCES FOR PLANT PEST AND DISEASE CONTROL (AS FORMULATED BY CODEX ALIMENTARIUS)

I. PLANT AND ANIMAL	
Preparations on basis of pyrethrins extracted from *Chrysanthemum cinerariaefolium*, containing possibly a synergist	Need recognized by the certification body or authority. Exclusion of Piperonyl butoxide after 2005 as a synergist.
Preparations of Rotenone from *Derris elliptica, Lonchocarpus, Thephrosia* spp.	Need recognized by the certification body or authority. The substance should be used in such a way as to prevent its flowing into waterways.
Preparations from *Quassia amara*	Need recognized by the certification body or authority.
Preparations from *Ryania speciosa*	Need recognized by the certification body or authority.
Commercial preparations/ products of Neem (Azadirachtin) from *Azadirachta indica*	Need recognized by the certification body or authority.
Propolis	Need recognized by the certification body or authority.
Plant and animal oils	
Seaweed, seaweed meal, seaweed extracts, sea salts and salty water	Need recognized by the certification body or authority. Not chemically treated.

contd. ...

... contd.

Gelatine	
Lecithin	Need recognized by the certification body or authority.
Casein	
Natural acids (e.g. vinegar)	Need recognized by the certification body or authority.
Fermented product from Aspergillus	
Extract from mushroom (Shiitake fungus)	Need recognized by certification body or authority
Extract from Chlorella	
Chitin nematicides	Natural origin
Natural plant preparations, excluding tobacco	Need recognized by certification body or authority.
Tobacco tea (except pure nicotine)	Need recognized by certification body or authority.
Sabadilla	
Beeswax	
II. MINERAL	
Copper in the form of copper hydroxide, copper oxychloride, (tribasic) copper sulphate, cuprous oxide, Bordeaux mixture and Burgundy mixture	Need, prescription and application rates recognized by certification body or authority. As a fungicide on condition that the substance be used in such a way as to minimize copper accumulation in the soil.
Sulphur	Need recognized by certification body or authority.
Mineral powders (stone meal, silicates)	
Diatomaceous earth	Need recognized by certification body or authority.
Silicates, clay (bentonite)	
Sodium silicate	
Sodium bicarbonate	
Potassium permanganate	Need recognized by certification body or authority.
Iron phosphates	As molluscicide.
Paraffin oil	Need recognized by certification body or authority.

contd. ...

... contd.

III. MICRO-ORGANISMS USED FOR BIOLOGICAL PEST CONTROLS	
Micro-organisms (bacteria, viruses, fungi) e.g. *Bacillus thuringiensis*, Granulosis virus,etc.	Need recognized by certification body or authority.
IV. OTHER	
Carbon dioxide and nitrogen gas	Need recognized by certification body or authority.
Potassium soap (soft soap)	
Ethyl alcohol	Need recognized by certification body or authority.
Homeopathic and Ayurvedic preparations	
Herbal and biodynamic preparations	
Sterilized insect males	Need recognized by certification body or authority.
Rodenticides	Products for pest control in livestock buildings and installations. Need recognized by certification body or authority.
V. TRAPS	
Pheromone preparations	
Preparations on the basis of metaldehyde containing a repellent to higher animal species and as far as applied in traps.	Need recognized by certification body or authority.
Mineral oils	Need recognized by the certification body or authority.
Mechanical control devices such as e.g., crop protection nets, spiral barriers, glue-coated plastic traps, sticky bands.	

SUBSTANCES FOR PLANT PEST AND DISEASE CONTROL (AS FORMULATED BY NPOP, INDIA)

Products for Plant Pest and Disease Control

Certain products are allowed for use in organic agriculture for the control of pests and diseases in plant production. Many of these products are restricted for use in organic production. Such products should only be used when absolutely necessary and should be chosen taking the environment impact into consideration.

Here "restricted" means that the conditions and the procedure for use shall be set by the certification programme.

I.	**Substances from plant and animal origin**	
•	*Azadirachta indica* [neem preparations (neem oil)]	Restricted
•	Preparation of rotenone from *Derris elliptica, Lonchocarpus, Thepphrosia spp.*	Restricted
•	Gelatine	Permitted
•	Propolis	Restricted
•	Plant based extracts (e.g. neem, garlic, pongamia, etc.)	Permitted
•	Preparation on basis of pyrethrins extracted from *Chrysanthemon cinerariaefolium,* containing possibly a synergist pyrethrum cinerafolium	Restricted
•	Preparation from Quassia amara	Restricted
•	Release of parasite predators of insect pests	Restricted
•	Preparation from Ryania species	Restricted
•	Tobacco tea	Not allowed
•	Lecithin	Restricted
•	Casein	Permitted
•	Sea weeds, sea weed meal, sea weed extracts, sea salt and salty water	Restricted
•	Extract from mushroom (Shiitake fungus)	Permitted
•	Extract from Chlorella	Permitted
•	Fermented product from Aspergillus	Restricted
•	Natural acids (vinegar)	Restricted
II.	**Minerals**	
•	Chloride of lime/soda	Restricted
•	Clay (e.g. bentonite, perlite, vermiculite, zeolite)	Permitted
•	Copper salts/inorganic salts (Bordeaus mix, copper hydroxide, copper oxychloride) used as a fungicide, maximum 8 kg per ha per year depending upon the crop and under the supervision of inspection and certification agency	Restricted
•	Mineral powders (stone meal, silicates)	Not allowed
•	Diatomaceous earth	Restricted
•	Light mineral oils	Restricted
•	Permaganate of potash	Restricted
•	Lime sulphur (calcium polysulphide)	Restricted
•	Silicates (sodium silicate, quartz)	Restricted
•	Sodium bicarbonate	Permitted
•	Sulphur (as a fungicide, acaricide, repellent)	Restricted

contd. ..

... contd.

III.	**Microorganisms/Biocontrol agents**	
•	Viral preparations (e.g. Granulosis viruses, Nuclear polyhydrosis, viruses etc.)	Permitted
•	Fungal preparations (e.g. Trichoderma species etc.)	Permitted
•	Bacterial preparations (e.g. Bacillus species etc.)	Permitted
•	Parasites, predators and sterilized insects	Permitted
IV.	**Others**	
•	Carbon dioxide and nitrogen gas	Restricted
•	Soft soap (potassium soap)	Permitted
•	Ethyl alcohol	Not allowed
•	Homeopathic and Ayurvedic preparations	Permitted
•	Herbal and biodynamic preparations	Permitted
V.	**Traps**	
•	Physical methods (e.g. chromatic traps, mechanical traps, light traps, sticky traps and pheromones)	Permitted
•	Mulches, nets	Permitted

REFERENCES

Handbook of Agriculture, 5th Edition, ISBN: 81-7164-050-8

"Integrated Pest Management – concepts and approaches" – Dhaliwal and Arora, Reprint Edition 2004, ISBN 81-7663-904-4

Websites

http://agritech.tnau.ac.in/

http://india.gov.in

http://www.oneplan.org

http://www.organicfarmingworld.com

Insect pest management for organic crops - ISBN 978-1-60107-042-5

Official website of Codex Alimentarius

Official website of FAO

Official website of NPOP, India

❑❑❑

Chapter **9**

Kunapjala (Organic Manure and Pesticide) for Crop Health

S.L. Choudhary and Sunil Khandelwal

AAHF Rajasthan Chapter, 105, Vidyanagar, Hiran Magri, Sect.4, Udaipur, Rajasthan

There are two basic requirements for production of healthy crops or we may say that proper nourishment and plant protection are the two important component for crop health management.

In Vrikshayurveda of Surapala about one thousand ago, kunap has been mentioned as liquid manure. *Kunap* is a Sanskrit word derived from the word *'Ku'* meaning dirty and *'nap'* means water. In verses 101 to 105 Surapala describes; "The excreta of marrow of bones, and flesh, brain and blood of a boar mixed with water and stored underground is called *kunap*. As per availability the fat, the marrow and flesh of the fish, the ram, the goat and other horned animals should be collected and stored. These should be boiled after mixing with water, and the mixture should be stored in an oilpot after adding sufficient quantity of husk (source not mentioned but probably it is paddy). After roasting in an ironpot sesame oil cake and honey should be added. Soaked black gram of good quality should also be added. A little ghee (butter milk) should then be poured. The items stated above should be taken at random as there is no measure for anything. One by one it should be placed in the pot in a warm place by a competent person.

Surapala further quoted that "this *kunapjala* is highly nourishing for trees. This is stated by the ancient sages and I (Surapala) repeat it here after verifying the same".

Simillarly the verses 171-174 of Sarangadhara's Upavanvinoda (1283-1301 AD) and Chakrapani Mishra's Vishvavallabh mentioned about *kunapjala* and its constituents described by Sarangadhara and Chakrapani Mishra are almost the same with slight variation. Sarangadhara also indicated that any material waste can be used for the preparation of *kunapjalal*.

Since the components of *kunapjala* mentioned by these scholars may not be easily available and every farmer can not handle them as this *kunap* is a non-vegetarian *kunap*, therefore, looking towards availability of nutrients and antimicrobial compounds as reported in other wasted products, efforts have been made to develop vegetarian *kunap* and simple non-vegetarian *kunap* for nourishment and protection of crops. Some of such preparations which have been tried over and further to be tried are as follows:

HERBAL *KUNAPJALA*

This is prepared from plants, which are either considered unwanted or called weeds in the modern agriculture and are destroyed under weed management systems. Some of them are having anti-microbial properties also. The materials required for preparation of this *kunapajala* are: plastic container of 200 L capacity (with lid), raw cow dung- 10 kg, fresh chopped leaves (locally available weed plants or other plants)-20 kg, molasses/cane jaggery-2kg, sprouted black gram seeds-2kg, water 10 L., and water-around 100 L.

Method of Preparation

1. Fresh and green leaves of weeds or other plants are chopped and kept in the plastic container.
2. Cow dung, sprouted gram seeds, and molasses/cane jaggery are added
3. Water is then added and the mixture is stirred well.
4. The lid is then closed
5. The container is kept in a warm place for 10-15 days depending on the temperature.
6. The mixture is stirred clockwise and anticlockwise 2-3 times daily to release the gas formed.
7. The preparation is ready to use when it stops producing gas.
8. It is filtered and then used.
9. The solution is used at 1 to 2% concentration or even in pure form depending on the condition of the plants.

DHANYAGAVYA

Dhanyagavya is made by fermenting paddy husk in cow urine or cow dung and water at least for one month. To prepare *dhanyagavya*, put 40 kg of paddy husk in 200 L plastic container, then add 20 kg of cow dung or 40 L cow urine and filled the drum up to the brim and covered with the lid. The mixture is stirred well 2-3 times a day in clockwise and anticlockwise direction. The end product contains silica in soluble form which can be sprayed and drenched in soil at required concentration for the control of fungal diseases.

INDASAFARI

As per Vrikshayurveda, the animal products are boiled for the *kunapjala*. The indasafari is a *kunapjala* prepared by using dead fishes which are discarded by the fisherman as well as waste of fishes. In this preparation the waste fishes with cow dung and water are fermented for 10-15 days. Some quantity of molasses or cane jaggery may be added to avoid the foul smell emit out from the preparation during stirring. This preparation can be used as drench as well as foliar spray with functions both as growth promoter and as an insecticide. This is quite effective to destroy the termites besides other pests.

COW DUNG WATER TO PROTECT FROM DROUGHT

Now a days drought is a severe problem throughout the country due to unusual climate especially eratic rainfall during monsoon. Spraying of cow dung water in such condition was found effective to save crops from dehydration up to 30-40days. The materials required for this preparation are: cow dung, rice water (starch), molasses/cane jaggery, pastic barrel of 200 L capacity, nylon bag, and water.

Method of Preparation

1. Take 30 kg fresh cow dung, 2 kg molasses/cane jaggery and 4 L rice water (starch).
2. Tied the above material in a nylon bag (50 kg capacity.)
3. Suspend this bag in a 200 L barrel filled with water.
4. Allow it to ferment for 36-48 h in this condition.
5. The preparation is ready to use after 48 h.

Spray this solution at 10% concentration to save the crops from drought.

❑❑❑

Chapter **10**

Biofertilizer : Its Production and Application in Organic Farming

S.K. Rathi

Production Manager, NAFED Biofertilizer, SPL-80, RIICO Industrial, Bharatpur-321 001

BIOFERTILIZER

Any biological organism living or dead used for the purpose of enriching soil fertility is known as Biofertilizer, or these are the beneficial organism used for:-

- N_2 enrichment - Example: *Rhizobium, Azotobacter, Azosprillum Acetobacter, Azolla, Cynobacteria, frankia*
- P solubilization - Ex.: *Pseudomonas, Fusarium, Aspergiillus*
- P mobilization - Ex.: *Mycorrhizae*
- Providing plant growth promoting substances like sedrophores,growth hormones, antimicrobial substances
- Ex.: *Azotobacter, Pseudomonas.*

CLASSIFICATION OF BIOFERTILIZERS

Biofertilizers can be classified in to mainly three catagories.

1. Nitrogen Fixers
2. Phosphate solubilizers and mobilizers
3. Organic matter decomposers

Some of the organisms which are part of biofertlizers also produce growth promoting substances, provide protection against drought and soil borne diseases. Commercially produced biofertilizers play an important role in modern agriculture as chemical fertilizer are limited and injurious to environment, soil and living organisms. Being low cost inputs these can be incorporated with chemical fertilizers and even can replace these in case of organic farming. In India carrier based biofertilizers have been produced since long time and now some units have started production of liquid biofertilizers. Biofertilizers consist of active part which has different microorganisms isolated and multiplied in laboratory through a suitable nutrient medium. This medium along with multiplied microorganism is mixed with another compotent called carrier. Carrier is the bulky component of biofertilizer which perseve the microorganisms active till they are applied. Microorganisms in their original nutrient medium are too concentrated to be applied uniformly. Carrier dilute this concentration and make uniform application possible. Carriar also provide sufficient nutrient to microorganisms for their survival.

History—If we go back into the history, the practice of inoculating seeds with culture dates from 1887. In India commercial production and use of Biofertilizer started in later half of nineteenth century . The positive approach to set the things in proper fashion came in early eighties of nineteenth century, where some of establishments and leading institutional agencies like Nafed were entrusted with job of production and promotion of biofertilizers. The next major break came in 7th plan in which the National Biofertilizer Development Project was established. Nafed, an apex level cooperative organisation initiated the commercial production and marketing of biofertilizer in the year 1985. Later on some other government and private organization also started production of biofertilizers.

PRODUCTION OF BIOFERTILIZER

Isolation /Procurement and maintenance of strains—The strains which are used for commercial production are isolated from various locations regularly and tested for their efficiency like nitrogen fixation, phosphate solubilization etc. Some efficient strains recommended for commercial production are obtained from different universities and research Institutions like GBPUA&T, HAU, IARI, ICRISAT, NCOF etc. Strains are revived quarterly and passed through the plants annually. A large number of strains of biofertilizers like *Rhizobium, Azotobacter, Azospirillum*, PSM, BGA, *Micorrhizae* etc., covering almost all the crops have been isolated and tested for efficiency. Before the onset of every season the strains are tested for their efficiency in plant growth room in the case of every strain which has to be mass multiplied ensuring the good quality and efficient biofertilizer.

Strains selection—For the selection of good strains exercise is done at growth chamber and micro-plots, the criteria used for selection of strain are (1) the strain should be able to fix nitrogen or solubilize phosphate with all the hosts for which the culture is recommended (2) it should have an ability to possess traits under wide range of field conditions (3) it should survive in soil saprophyticaly in the absence of host.

Preparation of starter cultures—The 250 ml. conical flasks containing 75ml. required liquid medium are sterilized in autoclave. These are inoculated with required strains in aseptic conditions. The flask are allowed to be shaken on rotary shaker at the temperature required by the strain. After a certain duration, it may be 3 to 9 days, the broth is checked for purity and proper growth. Again the same procedure is used for preparation of 1000 ml. conical flasks and these flasks are inoculated with the culture in small flask, same conditions are given for growth and again checked for proper growth and purity. Computer or coloured TV attached with video camera can be used for instant check up of purity and homogenity of the culture.

Fermenter stage—For mass production of biofertilizer fermenters are used. Same medium is prepared in seed fermenter having working capacity of 75 lit. and sterilized with the help of boiler. After cooling , the sterilized broth is inoculated with 5% inoculum which was prepared in big flasks. Like this main fermenters are charged and inoculated with 5% inoculum present in seed fermenters aseptically. After 48 hours we have mature inoculum which is mixed with carrier. Fermenters should be equipped with all the modern facilities like top opening for controlled air inlet and air outlet attached with filters and moisture separaters, inoculation port, digital meters showing temperature and pressure, sampling cock at the bottom, jacket for maintaining temperature and cooling of broth, outlet for cleaning the fermenters, to favour solution of oxygen, the air bubbles should be made as small as possible by using impellers, pipe line for transferring the broth to blender. The quality of the broth is checked aseptically at different intervals, when required population is gained, broth is transferred to blender.

Mixing of mature broth with the carrier—The mature broth is mixed with the milled (200mesh) dried lignite powder or other carrier to provide a moisture contents between 30 to 40% in a modernized blender. The pH is adjusted with the help of calcium carbonate. The moist carrier is sieved through a course sieve to remove lumps and then cured in bags for four days at room temperature for different reasons like for multiplication of organism acclimitization to the carrier, dissipation of heat of wetting.

Packing—After curing, the mixed material is manually or automatically packed in 50 to 75 micron low density milky white polythenes packets according to BIS norms. This provide an excellent barrier to water loss. These packets are sealed by electric sealers leaving 2/3 rd vacant space to give proper aeration to the inoculants. LDPE bags are also suitable for gamma irradiation.

Labelling—The biofetilizer packets are labeled with relevant information in respect of culture which is useful for end user. Packets should contain following information.

(1) Name of host for which content are meant
(2) Area of land quantity for which content are suitable
(3) Manufacturing & expiry date
(4) Batch no.
(5) Viable count
(6) Name of carrier used
(7) Instruction for storage
(8) Address of manufacturer

Outer Packing and storage of biofertilizer packets—Pin holes are made for proper aeration, these packets are again packed in seven ply semi virgin corrugated boxes, some packet are randomally checked for organism population by plate dilution method and further its performance at field level. The finished products are stored in cold room having temperature around 7ºC and from there, these are supplied to the farmers.

Precaution to be taken by the manufacturers—It is essential for the producers to adopt rigid quality control of the product, manufacturer should ensure that strain meant for nitrogen fixation, phosphate solubilization/mobilization decomposition, secretion of plant growth promoting substance is performing its work properly in the trials also. Total count of organism is proper in the liquid medium and finished products. Shelf life of the product is in accordance with the claims and is working efficiently during the shelf life. Standardized method should be adopted during production, packing and storage of biofertilizers. Distribution of the biofertilzers should be done quickly and these should not be exposed to high temperature.

Pecautions to be taken by the farmers—Apart from precaution to be taken by the manufacturers, certain precautions should be taken by the farmers also like 1.Store the biofertilizers at cool and shaded place 2.Keep them away from direct sunlight and heat 3.Avoid contact of biofertilizers with insecticides & pesticides 4.Expiry date is mentioned on each biofertilizer packet, it should be used before the expiry date 5.Different methods of application have been suggested such as Seed treatment, Seedling treatment, Soil treatment, Top dressing etc., required method should be applied.

Constraints and bottle neck in biofertilizer industry—At present India faces many constrains in the field of production, distribution and marketing of biofertilizer. Which can be divided in to physical, chemical, biological and technical constrains. In the summer, some parts of India have high temperature it is not suitable for biofertilizer storage, transportation and use. Highly acidic or saline soils also adversely effect population of introduced organisms. Bacteriophages, protozoa and nematodes act as predators of biofertilizers added to soil. Good quality peat and lignite are only available in Nilgiri and Neyveli resulting in high cost of transportation. Suitable strains for popular crops are available but these do not cover all the crops . Sterilization of carrier is not complete, only gamma irradiation is proper sterilization which can not be afforded by all procedure due to its high cost. Biofertilizers contain 30 to 40% moisture and can not be packed by a form, fill and seal machine. Uniform quantity of biofertilizer is not recommended through out India, recommendation varies form 75 gm to 500gm.People are not aware of governments policies related to biofertilizers, more training and publicity is required. Field response of biofertilizer is not immediately observed as compared to chemical fertilizers. Response is also low due to unfavorable soil conditions, temperature, pH., water logging, drought, deficiency of nutrients etc.

TYPES AND APPLICATION OF BIOFERTILIZER IN ORGANIC FARMING

RHIZOBIUM—*Rhizobium* bacteria are known for fixing atmospheric nitrogen since long. Nitrogen being the most important plant nutrient for increasing crop productivity, extensive studies were conducted by different scientists which proved that nodules formed by *Rhizobium* on legume roots are responsible for fixing atmospheric nitrogen

Specifications Mainly two genus are found 1. *Rhizobium*- fast growing 2. *Bradyrhizobium*- slow growing . Bacteria are small to medium in size, motile when young, gram negative, rod shaped,aerobic, grow well on CRYEMA (Congo Red Yeast Extract, Mannitol Agar) medium

Doses and Method of Application—

Seed Treatment—200gm to 500 gm powder based *Rhizobium* is mixed with required quantity of water seeds to be sown in one acre of land are coated with this slurry, shade dried and seeds are sown immediately.

Soil Treatment—Two to three kg of *Rhizobium* inoculant is mixed with 50kg of mature compost or cowdung and kept for overnight, the mixture is spread in one acre land before last ploughing.

Benefited Crops—All leguminous crops.

Benefits—*Rhizobium* inoculants can increase crop yield by 30% and 40 to 80 kg. per hectare residual nitrogen is left for next crop.

AZOTOBACTER—*Azotobacter* bacteria have multiple action, they can fix atmospheric nitrogen in non-leguminous plants, it also release growth promoting substances like IAA, Gibberellins, Vitamins and Ascobic Acid. *Azotobacter* can control plant pathogens, also known to slubilize phosphorus.

Specifications—Mainly six species have been isolated but *Azotobacter chroococcum, azotobacter vinelandii* and *azotobacter beijerinckii* are much popular. It is free living gram negative, aerobic bacterium, can be isolated on nitrogen free medium, cells are polymorphic, in young stage have flagella, rod shaped to coccal in old cells, measure (2-10 μ). *Azotobacter* are known to form cyst to which stand adverse conditions

Doses and Method of application—

Seed treatment—400gm to 1000gm powder based *Azotobacter* is mixed with sufficient quantity of water, seeds to be sown in one acre of land are coated with this mixture. After shade dry seeds are sown immediately.

Cutting and Seedlings Treatment—1200 gm powder based *Azotobacter* is dissolved in 20 to 30 lit water. Cutting and seedling are dipped in this slurry for five minutes and these are sown immediately

Soil Treatment—Two to three kg of *Azotobacter* powder is mixed with 50 to 100 kg. of mature compost before last ploughing.

Benefited Crops—All non leguminous crops including cereal and horticultural crops

Benefits—Its application can increase 20% crop yield apart from reducing plant diseases and providing plant growth promoting substances.

AZOSPIRILLUM—*Azoto* means nitrogen and *spira* means spiral. *Azospirillum* is a free living associative nitrogen fixing bacterium. It fixes nitrogen at low partial pressure of oxygen. Also known to produce growth promoting substances like Indole Acitic Acid, Gibberelins, Cytokinins.

Specifications—Two groups namely *Azosprillum lipoferum* and *Azospiriullum brasilense* are popular. It can grow well on semi soild nitrogen-free medium, cells gererally are vibroid but may be S-shaped or helical, diameter is about 1.0 μ, single polar flagellum in liquid and lateral flagella on solid medium, gram negative, aerobic bacteria.

Doses and Method of Application—

Seed Treatment—400 gm. to 1000 gm. powder based *Azospirillum* is dissolved in required quantity of water to coat seeds to be sown in one acre land, after shade dry seeds are sown immediately.

Cutting and Seedling Treatment—1200 gm. *Azospirillum* powder is mixed with 20 to 30 lit. of water cutting and seedlings are kept for five minutes in this mixture, are sown immediately.

Soil Treatment—Two to three kg of powder based *Azospirillum* is added to 50kg of mature compost and is broadcasted in one acre of land prior to last plounghing.

Benefited Crops—Sorghum, pearl millet, finger millet, maize, wheat, rice, few vegetables, mulberry etc.

Benifits—It can increase crops yield from 5% to 10 % also improve over all growth of plants.

ACETOBACTER—*Acetobacter diazotrophicus bacterium* can fix atmospheric nitrogen in high sugar containing plants very efficiently. The bacteria multiply around the roots and fix nitrogen. Proved to catter ¼ of total nitrogen requirement of plants.

Specifications—*Acetobacter* is an endophyic *diazotroph*. Bacteria are small, gram negative, rod shaped aerobic in nature. Form pellicle in N-free semi soild with 10% sucrose on LGI medium.

Doses and Method of Application—

Soil Treatment—Two to three kg of biofertilizer is mixed with 50 kg of mature compost which is broadcasted in one acre land. After ploughing the land seeds can be sown.

Benefited crops—*Acetobacter* can be used in plants containing high sugar like sugarcane, sugarbeats and coffee.

Benefits—We can save ¼th of total nitrogen requirement of plants by supplementing these bacteria. General health of plant is also improved.

FRANKIA—These are the organisms which leads to non–leguminous symbiosis. Almost 120 species of mainly trees and shrubs are known to contain nitrogen fixing organisms in their root nodules but these are not *rhizobia*. *Frankia* is collective name for actinomycetes which develop association with non lguminious plants.

Specifications—Nodules formed by *Frankia* in some cases contain haemoglobin and in others haemoglobin is absent but all have a system to protect nitrogenase. Hyphae pass through walls of host cells, parallel to cell wall or in the wall itself. Hyphae have cytoplasmic membrane and the host cell plasmellemna. Vesicles are generally formed by frankia excepting in *Casuarinas*, which indicate effectiveness.

Dose and Method of Application—

Soil Treatment—Two to three kg of *Frankia* can be used for mixing with 50 kg of mature compost which is broadcasted in one acre of land and after ploughing the land seeds are sown.

Benefited Trees—*Frankia* can be used in tress like *Alnus, Casuarina, Ceanothus, Eleagnus,Rubus Myrica, Hippophae, Pushia* etc.

Benefits—Nitrogen saving is from 12 to120 kg. per hactre per year. It is also used for upgrading poor soils.

AZOLLA—*Azolla* is a fresh water floating fern, out of six spices four are popular. *A nilotica, A maxicana, A microphylla* and *A pinnata* are common in India. *Azolla* have symbiotic association with *Anabaena azollae* (Cynobacterium) and fix appreciable amount of nitrogen in the rice field. The association utilize the energy from photosynthsis to fix atmospheric nitrogen.

Specifications—Consists of main rhizome and length is 3 to 4 cm. Upper side of stem and central cavity shelter *Anabaena azollae,* heterocyst part of this organism is site of nitrogen fixation. Sexual reproduction takes place but usual mode of multiplication is not there forms a green mat on water, later become reddish due to accumulation of anthocyanin pigments. Temperature, light intensity and phosphosrus contents play a major role In functioning of the system.

Doses and Method of Application—

Broadcast—200 kg Azolla is sufficient for broadcasting over 7 to 10 days rice plantation in one hectare of land, mat is formed which is incorporated in soil.

Soil Treatment 10 M.T. fresh *Azolla* is applied basally to one hectare of land and mixed properly by ploughing.

Benefited Crops—Most benefited crop is rice.

Benefits—We can save 40 -60 kg. of N/Ha and increase in yield is 15-20%. It can also be used as green manure because of its large biomass and high nitrogen content in flooded rice.

BLUE GREEN ALGAE (CYNOBACTERIA)—Algae participating are *Anabaena, Nostoc, Calothrix, Tolypothrix, Westialla, Slgonema* etc. They form non fibrous slimy growth on the ground, on the surface of soil and water. They have oxygen evolving photosynthesis and can survive in extreme climatic conditions. Cynobacteria fix nitrogen and excrete vitamin B12, auxins, ascorbic acid which contribute to growth of rice.

Specifications—Cynobactria are prokaryotic, photolithotrophic microscopic plants. Fixation of nitrogen takes place in specialized cells called hetrocysts present on the algal filament. Hetrocysts are thick walled, large apparently empty cell growing in between pigmented cells. They have a sexual reproduction and form motile stages, their food reserve is glycogen.

Doses and Method of Application—

Soil Treatment—Five kg. of composit inoculants of blue green algae is used to inoculate one acre of land, the field is flooded with water to level 2-5cm. followed by broadcasting of 2 kg. of superphosphate. Algal inocula should be applied for two to three consecutive seasons.

Benefited Crops—Inoculation is very useful in rice crops.

Benefits—Crops yield can be increased by 10 to 15 % and 30 kg. N/Ha can be saved. Algal inoculation is environmental friendly, add organic matter to soil, excrate growth promoting substances and improve soil properties.

PHOSPAHTE SOLUBILIZING BACTERIA/MICRORGANISMS—These are the organisms which solubilize unavailable phosphorus (fixed form) and make it available to plants. In this available form is HPO_4 or PO_4. They produce organic acid/ inorganic acid and bring about solubilization. Many soil bacteria belonging to genera *Pseudomonas* and *Bacillus,* fungi belong to genera *Penicillium* and *Aspergillus* have the ability to bring insoluble phosphate in soil into soluble forms by secreting organic acid which lower the pH and bring about dissolution of bound forms of phosphate. Mainly participating organism are *Pseudomonas striata, Pseudomonas sp. Bacillus polymyxa, Bacillus circulans, B. subtilis, B. magatrium, Aspregillus awamori, Penicillium digitatum, Fusarium oxysporum* etc.

Specifications—Microorganism produce organic acid such as lactic glycolic, succinic, lactic, acetic, fumeric,citric etc. Which lead to solubilization of phosphate some hydroxy acids chelate with calcium and iron resulting in effective solubilization and utilization of phosphates. All phosphate solubilizing microorganisms can be cultured on Pikovaskaya's medium. Most efficiently working, *Pseudomonas* bacteria are gram negative, non spore forming, non capsulating and motile rod shaped.

Doses and Method of Application—

Seed Treatment—200 gm. to 500 gm. powder based culture is added to sufficient quantity of water so that seed required for one acre of land can be coated, after shade day seeds are sown immediately.

Cutting and Seedling Treatment—600 gm. powder based PSB (PSM) is dissolved in 15 to 20 lit. of water and cutting or roots of seedling are dipped for five minute in this solution before transplantation.

Soil Treatment—Two to three kg of powder PSB (PSM) is mixed with 50kg.of mature compost or cowdung. This mixture is broadcasted in one acre of land before last ploughing.

Benefited Crops—PSB (PSM) can be applied to any crop or plant.

Benefits—50% phosphate fertilizer can be saved by using PSB (PSM) biofertilizer, crop /plant remains healthy and 30 to 40% yield can be increased by its application.

MYCORRIZAE—*Micorrahizae* word was coined by German Botanist Frank in the year 1885. It is symbiosis between fungi and roots of plants. Three type of *mycorrizae* are mainly found.

Ectomycorrihiza- has compact fungus mantle due to thickened branched fungi over the root surface, and inter celllar hyphae in the cortex, mostly found in forest trees.

Endomycorrhiza- has loose net work of hyphae in the soil surrounding the root, with the extensive hyphal growth within the root cortex present in most of the angiosperms.

Ectendomycorrhza- is similar to *enctomycorrhiza* type but has both inter and intra cellular hyphae, these are also associated with forest trees.

Specifications—Here we are mainly concerned with *endomycorrhizae* particularly with vesiculer arbusculer *mycorrhiza* popularly known as VAM or AM. In this terminal swellings called vesicles and branching known as arbuscules are formed, fungi participating are *Glomus, Gigaspora, Acaulospora, Sclerocystis, Entrospora* and *Scutelospora*. Due to extensive absorption of phosphate by the root a phosphate deficit zone is created around the root. The hyphae of these fungi cross this phosphate deficit zone and absorb minerals from newly explored area The hyphae function as root hairs of the plant. Vesicles formed at the end of the hyphae are storage organs, usually contain phosphate which is used by the plant. Arbuscules transfer the nutrients to the plant.

Doses and Method of Application—*Mycorrhiza* inoculum is soil based, consist of hyphae and spores.Inoculum is applied in following ways.

Hole Method—Make a small hole for putting seed in soil, place soil based inoculum, put some soil over inoculum and then put the seeds, cover with soil.

Layer Method—Take out a layer of soil spread inoculum properly, put some soil and then place the seeds and cover with removed soil.

Mix Method—Dig the soil,take it out and mix with inoculum, place it back and sow the seeds as usual.

In case of flyash or other carrier 5kg. inoculum per acre is to be used and soil treatment method can be adopted.

Benefited crops—*Mycorrhizae* biofertilizer can be used in any crop or plant barring mustard where response is low.

Benefits—*Mycorrhizae* provide immobile elements to the plants like phosphorus,copper and zinc. Enhance water transport,assist in rooting and survival of cuttings increaes tolerance to high temperature and heavy metals, plants are less susceptible to attack of pathogens like *Fusarium, Rhizoctonia* & nematodes.

ORGANIC MATTER DECOMPOSERS—organic manure and compost are source of plants nutrients. Cow dung and various agricultural wastes carry all required nutritions for plants ,but due to complex form the nutrients are not available to the crop.The agricultural wastes are first decomposed through the activity of certain microorganisms.The composting microorganism are present in the nature but complete degradation of agri- wastes takes around one year because of low population and less potency of composting microorganism. composting culture do this work within two to three months.

Specifications—Organic matter decomposer or composting culture is balanced blend of highly efficient microorganism, these are mainly *Trichoderma, Penicillium Crysoginum, Aspergillus nigar* etc. Decomposers can be enriched with *Azotobacter*, PSB and 1% rock phosphate.

Doses and Method of Application—One kg. of composting culture is sufficient for decomposing one tractor trolly or one M.T. of agri waste within two to three month. One kg composting culture mixed with ten lit. clean water is blended with agri-wastes, reshuffled weekly ensuring that proper moisture is maintained till the end of decomposition.

Benefited Wastes—Composting culture can be used for decomposing cow dung, agri-wastes, city-waste, vegalable and fruit wastes.

Benefits—Humus substances like humic acid and fulvic acid are produced, have short and long term advantages when applied to soil. Energy is stored for growth of beneficial microorganisms. Supply nutrients and growth promoting substances to the plants.

TRICHODERMA—*Trichoderma* is used as biocontrol against different fungal diseases, it acts as repellent to these soil and seed born fungal plant pathogens. *Trichoderma* is reported to check diseases like stem rot, root rot, damping off wilt, blight, which are caused by pathogens such as *Fusarium, Pythium, Rhizoctonia, Macrophomina, Phytopthora, Verticillium* etc.

Specifications—Two species of *Trichoderma* are popular in india namely 1. *Trichoderma viride* 2. *Trichoderma harzianum. Trichoderma* can be produced on any locally available organic substrate like mellets, molasses, grain husk, cow dung etc. Survives very well in soap stone or talc powder, which is also used as carrier for commercial production. Soap stone based *Trichoderma* contains spores along with corboxy methyl cellulose.

Doses and Method of Application—

Seed Treatment—Four to five gm. of carrier based *Trichoderma* is dissolved in required quantity of water to coat one kg. of seeds, sowing is done immediately.

Cutting and Seedling Treatment—600 gm. *Trichoderma* powder is mixed with 15 to 20 lit.of water, cuttings or seedlings are dipped in this solution for 5 minutes before transplantation.

Soil Treatment—Two to three kg. of powder based *Trichoderma* is mixed properly mixed with 50 kg. of mature compost and spread over one acre of land before last ploughing.

Benefited Crops—Reported to control diseases in pulses, vegetables, oil seeds etc., but can be used in any crop.

Benefits—*Trichoderma* functions as effective biocontrol against variety of seed & soil pathogens.

❑❑❑

Chapter **11**

Plant Growth Promoting Rhizobacteria (PGPR) Associated with Plant Nutrition

B.K. Mishra, S.K. Khandelwal and G. Rajamani

Department of Molecular Biology & Biotechnology, Rajasthan College of Agriculture, Maharan Pratap University of Agriculture & Technology, Udaipur

The soil bacteria which flourish in the rhizosphere of plants stimulate plant growth by a plethora of mechanisms. These bacteria are collectively known as PGPR (plant growth promoting rhizobacteria). These modes of action include fixing N_2, increasing the availability of nutrients in the rhizosphere, positively influencing root growth and morphology, and promoting other beneficial plant–microbe symbioses. The combination of these modes of actions in PGPR is important for sustained availability of plant nutrients in the soil. There are many challenges before widespread utilization of PGPR for plant nutrition. The searches for PGPR are increasing at a rapid pace as efforts are made to exploit them commercially as a tool for sustainable agriculture.

Plant growth promoting rhizobacteria (PGPR) represent a wide variety of soil bacteria which, when grown in association with a host plant, result in stimulation of growth of their host. The term biofertilizer refers to the use of soil microorganisms to increase the availability and uptake of mineral nutrients for plants. The focus of chapter is the mode of action of PGPR which act as nutrient bio-suppliers, either directly by helping to provide nutrient to the host plant, or indirectly by positively influencing root growth and morphology or by aiding other beneficial symbiotic relationships. Not all PGPR are nutrient bio-suppliers. Many PGPR stimulate the growth of plants by helping to control pathogenic organism (as biocontrol agents).

Biofertilizer is as a substance which contains living microorganisms which, when applied to seed, plant surfaces, or soil, colonizes the rhizosphere or the interior of the plant and promotes growth by increasing the supply or availability of primary nutrients to the host plant. This definition is based on the logic that the term biofertilizer is a contraction of the term *biological fertilizer*. This definition separates biofertilizer from organic fertilizer (fertilizer containing organic compounds which directly, or by their decay, increase soil fertility). The rhizospheric microorganisms which improve utilization of soil nutrients, should be called nutrient bio-suppliers. Whether the existence of a microorganism increases the growth of plants by replacing soil nutrients (e.g., by biological N_2 fixation (BNF)) or making nutrients more available (e.g., by solubilization of phosphates) or increasing plant access to nutrients (e.g., by increasing root surface area), as long as the nutrient status of the plant has been enhanced by the microorganism, the substance that was applied to the plant or soil containing the microorganisms, is also called as biofertilizer. Not all PGPR can be considered nutrient bio-suppliers. Rhizospheric bacteria that promote plant growth by control of deleterious organism are called as biopesticides. Interestingly some PGPR appear to promote growth by acting as both biofertilizer and biopesticides. For example, strains of *Burkholderia cepacia* have been shown to have biocontrol characteristics to *Fusarium* spp., but also can stimulate growth of maize under iron-poor conditions via siderophore production . As defined, nutrient bio-suppliers must contain living microorganisms which promote plant growth by improving the nutrient status of the plant.

Relationships between PGPR and Plants—For PGPR to have a beneficial effect on plant growth via an enhancement of the nutrient status of their host, there obviously needs to be a intimate relationship between the PGPR and the host plant. However, the degree of intimacy between the PGPR appears to have and the host plant can vary depending on where and how the PGPR colonizes the host plant. Relationships between PGPR and their hosts can be categorized into two levels of complexity:

(1) rhizospheric

(2) endophytic.

In rhizospheric relationships, PGPR may colonize the rhizosphere, the surface of the root, or even superficial intercellular spaces. Plant changes the physical and chemical composition of the soil in the rhizosphere compared to the bulk soil which can affect the ability of a PGPR to colonize the rhizosphere. These differences are manifested by changes in soil pH, water potential, partial pressure of O2, and a myriad of other physical and chemical characteristics due to plant exudations. In many rhizospheric relationships, the PGPR will actually attach to the surface of the plant. Scanning electron micrographs of bacteria on the surface of plants roots are common in the scientific literature. However, the means of attachment is much less often known. *Azospirillum* sp. is one group of PGPR where mechanisms involved in attachment have been well characterised . Colonization of root surfaces by PGPR is not uniform. For example, *Kluyvera ascorbata* colonized the upper two-thirds of the surface of canola roots, but no bacteria were detected around root tips. In endophytic relationships, PGPR actually reside within apoplastic spaces inside the host plant. Although there is some evidence of endophytes occupying intracellular spaces, these reports are rare. The best characterized symbioses involving colonization of hosts by endophytes are the legume–rhizobia symbioses. The chemo-attraction, attachment, and infection of the microsymbiont and the development of the root nodules that house the bacteria is a highly complex and highly regulated process. Likewise the infection process and development of N_2-fixing specialized structures in the non-legume symbioses of

Parasponia–rhizobia, *Alnus–Frankia*), *Azolla– Anabaena, Gunnera–Nostoc* and cycads–cyanobacteria have also been well characterized. In many endophytic relationships where no specialize structures such a root nodules are formed, the means of infection of the PGPR and the location and environmental milieu in which the microsymbiont is found is not as clearly understood as in legume–rhizobia symbioses. Firstly, the means of infection is not clear. Although many studies exist which point to entry of PGPR into apoplastic spaces of the host at the junction of lateral roots from primary roots presents elegant arguments that much of the evidence for this form of entry is artefactual. Some PGPR endophytic species are known to have cellulose and pectinase activities and these activities could not doubt aid in infection, however heavily suberized and/or lignified cell layers would still present a barrier to such bacteria. Some endophytic PGPR may utilize other organisms as vectors to gain access to apoplastic spaces in their host. For example both the pink sugarcane mealybug (*Saccharicoccus sacchari*) and arbuscular mycorrhizae have been implicated in the infection of host plants by the endophytic diazotroph, *Gluconacetobacter diazotrophicus* (formerly *Acetobacter diazotrophiucs*). Depending on the host plant and the endophyte, nutrient bio-supplying PGPR may be found in all parts of plants (seed, roots, stems, leaves, fruit, etc). Within these organs, there is evidence of endophytes living in apoplastic intercellular space within parenchyma tissue and xylem vessel apoplast. In fact, while some researchers consider xylem apoplast an ideal place for the location of endophytes (for constant supply of nutrients; circulation systems for beneficial products from the microsymbiont) others feel that this is a most unsuitable habitat for endophytes, and note that many plant pathogens bring about plant death by colonization of the xylem apoplast.

MODE OF ACTION OF PGPR AS NUTRIENT BIO-SUPPLIERS

The means by which PGPR enhance the nutrient status of host plants can be categorized into five areas:

(1) Biological N_2 fixation,

(2) Increasing the availability of nutrients in the rhizosphere,

(3) Inducing increases in root surface area,

(4) Enhancing other beneficial symbioses of the host, and

PGPR THAT FIX N_2 FOR THEIR HOST

The most studied and longest exploited PGPR are the rhizobia (including the *Allorhizobium, Azorhizobium, Bradyrhizobium, Mesorhizobium, Rhizobium,* and *Sinorhizobium*) for their ability to fix N_2 in their legume hosts. Commercial rhizobia inoculants for use on legume crops were first introduced in the 1890s. It is interesting that so many PGPR have the ability to fix N_2, yet rarely is their mode of action for the stimulation of plant growth credited to BNF. PGPR that have the ability to fix N_2, but for which there is little evidence, or even counter evidence, that their stimulation of growth of a specific host plant is due to nitrogenase activity include *Azoarcus* sp., *Beijerinckia* sp., *Klebsiella pneumoniae, Pantoea agglomerans* and *Rhizobium* sp. It is also interesting that so many PGPR are diazotrophs, yet the mechanism underlying their growth promoting effects is not the supply of fixed N_2 to the host.

Despite much excitement and research on associative N_2 fixation in a vast array of non-legume crop plants there is still little evidence of inoculation of non-legumes leading to agronomically significant levels of BNF from PGPR in most crops. For example, it was once widely believed that the beneficial effects of *Azospirillum brasilense* on non-legumes was via BNF, but it is now well recognized that its growth promoting effects originate predominantly from other mechanisms (i.e., phytothormone production, effects on root morphology, etc.). An exception to the generally low influence of associative BNF are the symbioses between sugarcane and endophytic diazotrophs, in which the host routinely obtains 20–60% of its N requirements from the microsymbiont. In particular, *Gluconacetobacter diazotrophicus* has been shown to contribute significantly to the N-nutrition of sugarcane under controlled conditions. Another N_2-fixing endophytic species of sugarcane, *Herbaspirillum seroepdicae*, has been shown to infect rice and result in increased $^{15}N_2$ incorporation. Some researchers have been pursing the formation of 'para-nodules' on non-leguminous plants as potential sites to host diazotrophs. Treatment of the roots of some plants with natural, synthetic, or precursors of the phytohormone auxin will result in the formation of nodular-looking tumours with of varying degrees of tissue organization, a para-nodule. These tumours likely originate from lateral root meristems and are more correctly viewed as modified root tissue, rather than an organ with ontogenesis similar to legume nodules. Inoculation of such treated roots with rhizospheric or endophytic diazotrophs sometimes results in colonization of the para-nodule and some degree of nitrogenase activity. Such experiments have involved the colonization of para-nodules on associations involving wheat and various diazotrophs, *Azorhizobium caulinodans* and rice, *Azotobacter nigricans* and rape and *Azospirillum* sp. and maize but the formation of para-nodules offers little advantage in promoting associations between diazotrophs and non-legume hosts.

Table 1 : Plant growth promoting rhizobacteria as nutrient bio-suppliers

S.No.	PGPR	Relationship to host	Host crops
1.	*Azospirillum* sp.	Rhizospheric	Maize Rice Wheat
2.	*Azoarcus* sp.	Endophytic	Kallar grass Sorghum Rice
3.	*Azotobacter* sp	Rhizospheric	Maize Wheat
4.	*Bacillus polymyxa*	Rhizospheric	Wheat
5.	*Burkholderia* sp.	Endophytic	Rice
6.	*Anabaena* and *Nostoc*	Rhizospheric	Rice
7.	*Gluconacetobacter diazotrophicus*	Endophytic	Sorghum Sugarcane
8.	*Herbaspirillum* sp.	Endophytic	Rice Sorghum Sugarcane
9.	*Bacillus circulans* and *Cladosporium herbarum*	Rhizospheric	wheat
10.	*Enterobacter agglomerans*	Rhizospheric	tomato
11.	*Pseudomonas chlororaphis* and *P. putida*	Rhizospheric	soybean

PGPR THAT INCREASE THE AVAILABILITY OF NUTRIENTS IN THE RHIZOSPHERE

There is ample evidence that the mode of action of many PGPR is by increasing the availability of nutrients for the plant in the rhizosphere. The method by which these increases take place involve solubilization of unavailable forms of nutrients and/or siderophore production which helps to facilitate the transport of certain nutrients (notably ferric iron).

SOLUBILIZATION OF PHOSPHATES

Phosphorus is second only to nitrogen in mineral nutrients most commonly limiting the growth of terrestrial plants. Ironically, soils may have large reserves of total P, but the amounts available to plants is usually a tiny proportion of this total. The low availability of P to plants is because the vast majority of soil P is found in insoluble forms, and plants can only absorb P in two soluble forms, the monobasic (H_2PO_4) and the diabasic (HPO_4) ions. Phosphate solubilizing bacteria are common in the rhizosphere and secretion of organic acids and phosphatases are common method of facilitating the conversion of insoluble forms of P to plant-available forms. The solubilization of P in the rhizosphere is the most common mode of action implicated in PGPR that increase nutrient availability to host plants. Examples include *Azotobacter chroococcum* with wheat, *Bacillus circulans* and *Cladosporium herbarum* with wheat, *Enterobacter agglomerans* with tomato, *Pseudomonas chlororaphis* and *P. putida* with soybean etc. Phosphate-solubilizing bacteria are common in rhizospheres. However, not all P-solubilizing PGPR increase plant growth by increasing P availability to the hosts.

FACILITATED ABSORPTION OF IRON

Iron is an essential nutrient of plants, but it is relatively insoluble in soil solutions. Plant roots prefer to absorb iron as the more reduced ferrous (Fe^{2+}) ion, but the ferric (Fe^{3+}) ion is more common in well aerated soil although it is easily precipitated in iron-oxide forms (Salisbury and Ross, 1992). Plants commonly excrete soluble organic compounds (chelators and phytosiderophores) which binds Fe^{3+} and helps to maintain it in solution. Chelators 'deliver' the Fe^{3+} to the root surface where it is reduced to Fe^{2+} and immediately absorbed. Phytosiderophores, excreted by grasses, are absorbed with the Fe^{3+} across the plasmalemma. Some rhizospheric bacteria also produce siderophores and there is evidence that a number of plant species can absorb bacterial Fe^{3+}-siderophore complexes. However, there is a controversy as the significance of bacterial Fe^{3+}- siderophore uptake to the iron nutrition of plants. The contribution of these siderophores to the overall iron requirements of plants is small, but play an important role in calcareous soils. In fact, the vast majority of microbial siderophores in the rhizosphere is associated with their Biocontrol activities due to their competitive effects with plant pathogens.

POSITIVE EFFECTS ON ROOT GROWTH AND MORPHOLOGY

Root morphology parameters are very important in the uptake of a variety of nutrients (NO_3^-, P, and K). Threfore, PGPR that affect root morphology, and more specifically, increase root surface area, have a huge influence on nutrient uptake potentials. By far the most evidence for the positive effects of nutrient bio-supplying-PGPR points to bacteria-mediated changes in root growth and morphology. Bacterial mediated increases in root weight are commonly reported responses to PGPR inoculations. Many actual and putative nutrient bio-supplying-PGPR produce phytohormones that are believed to be related to their ability to stimulate plant growth. In most cases these phytohormones are believed to be changing assimilate partitioning patterns in plants and affecting growth patterns in roots to result in bigger roots, more branched roots, and/or roots with greater surface area. Indole-3-acetic acid is a phytohormone which is known to be involved in root initiation, cell division, and cell enlargement. This hormone is very commonly produced by PGPR. Most commonly,IAA-producing PGPR are believed to increase root growth and root length, resulting in greater root surface area which enables the plant to access more nutrients from soil. Production of other phytohormones by nutrient bio-supplying-PGPR has been identified, but not nearly to the same extent as bacteria which produce IAA. Cytokinins are a class of phytohormones which are known to promote cell divisions, cell enlargement, and tissue expansion in certain plant parts. Researchers have recently begun to identify cytokinin production by PGPR. Gibberellins (gibberellic acid; GA) are a class of phytohormones most commonly associated with modifying plant morphology by the extension of plant tissue, particularly stem tissue. Inoculation of alder (*Alnus glutinosa*) with these PGPR could effectively reverse a chemically induced inhibition of stem growth. Although this is not strong evidence of GA production being a common method of growth promotion by PGPR, it does suggest that it may have a role, and indicates that more research in this area is warranted. Ethylene is the only gaseous phytohormone. It is also known as the 'wounding hormone' because its production in the plant can be induced by physical or chemical perturbation of plant tissues . Among its myriad of effects on plant growth and development, ethylene production can cause an inhibition of root growth. It is believed that the mode of action of some PGPR is the production of 1-minocyclopropane-1-carboxylate (ACC) deaminase, an enzyme which could cleave ACC, the immediate precursor to ethylene in the biosynthetic pathway for ethylene in plants. ACC deaminase activity would decrease ethylene production in the roots of host plants and result in root lengthening. In some cases, the growth promotion effects of ACC deaminase-producting PGPR appear to be best expressed in stressful situations such as flooded or heavy metal-contaminated soils. Aside from the effects of bacterial-facilitated effects of phytohormones on root growth, there is evidence in some systems that PGPR may be directly affecting root respiration which in turn leads to increases in root growth. Inoculation of various plant species with *Azospirillum* has shown to increase root respiration rates. This increase in root respiration caused a compensatory increase in carbon assimilation in the shoot, leading to 8% increase in plant growth.

'HELPER' BACTERIA

It is not uncommon to see the effects of nutrient bio-supplying-PGPR via synergism or promotion of the beneficial effects of a third-party rhizospheric microorganism. In these cases, the PGPR aiding the other host–symbiont relationship is often referred to as 'helper'bacteria. The vast majority of studies investigating PGPR as aids of other host-symbiont relationships involve either legume–rhizobia symbioses or plant–fungi symbioses.

STIMULATION OF LEGUME–RHIZOBIA SYMBIOSES

There is evidence for a number of modes of action for PGPR stimulation of legume–rhizobia symbioses, but the most commonly implicated mode is phytohormone-induced (usually IAA) stimulations of root growth . In this way, the stimulation of nodulation is most commonly an indirect effect; the PGPR stimulates root growth, which provides more sites for infection and nodulation. Some PGPR that stimulate legume–rhizobia symbioses appear to more directly influence the development of the symbioses. They might stimulate root nodulation by mediating increased production of flavonoids by the legume host. These flavonoids are the initial chemical signals secreted by the legume host to induce *nod* genes in rhizobia and thereby initiate the legume–rhizobia symbiosis.

STIMULATION OF PLANT-FUNGAL SYMBIOSES

PGPR sometimes enhance plant growth indirectly by stimulating the relationship between the host plant and beneficial rhizospheric fungi such as arbuscular mycorrhizae (AM). Although AM by themselves are well known to enhance the uptake of various soil nutrients (especially phosphorus), studies are proving that co-inoculation with some PGPR can enhance the relationship between plant and fungal symbionts. The combination of the arbuscular mycorrhizal fungus *Glomus aggregatum*, and the PGPR *Bacillus polymyxa* and *Azospirillum brasilense* maximized biomass and P content of the aromatic grass palmarosa (*Cymbopogon martinii*) when grown with an insoluble source of inorganic phosphate. It is interesting that the bacteria are acting in concert with the AM to improve P acquisition of the host plant

CONCLUSIONS

There is huge potential for the use of PGPR as nutrient bio-supplying agents for a wide variety of crop plants in a wide range of climatic and edaphic conditions. The mode of action of nutrient bio-supplying PGPR indicate bacterial influences on plant growth regulating substances. These factors often affect root development and morphology resulting in greater root surface area for the absorption of nutrients or enhance other host–symbiont relationships (i.e., the 'helper' bacteria effect). Second. It effects on plant growth regulation, solubilization of phosphorus is the most commonly suggested mode of action of nutrient bio-supplying PGPR. Although, use of rhizobial inoculants on legumes existed for over a century there has been little commercialization of nutrient bio-supplying PGPR. No doubt the lack of consistent responses in different host cultivars and different field sites are reasons limiting the more widespread commercialization of nutrient bio-supplying PGPR.

❑❑❑

Chapter 12

Mycorrhizal Inoculation of Potential Medicinal Plants : Field Trials and Effect on the Bioactive Phytoconstituents

YUDHVIR K. BHOOM

Sri Venkateswara College (Uni. of Delhi), Dhaula Kuan, New Delhi-110021

Ayurved, the Science of Life is our ancient system of Medicine which has been preserved with us since Vedic Period and is now recognized globally as the most ancient and yet Scientific system. There is the global acceptability of Herbal Health Care Systems because of the least side effects. All these Traditional Systems of Medicines practiced in different countries in the world are all based upon the use of the medicinal plants which are found to grow in the Nature in the Forests and these medicinal plants have been exploited , harvested unscientifically and unjudiously from their Natural Habitats because of their heavy demands in the International Markets. Many of these plants have now been placed in the RET category. Efforts have been initiated to undertake the ex-situ and insitu conservation of these Medicinal Plants by National Medicinal Plants Board, Deptt. Of AYUSH, MHFW, GOI right from 2002 and also by promoting the Cultivation of the Medicinal Plants since 2002 by the Farmers in different states . The cultural practices of the prioritized 32 Medicinal plants have also been developed and are being used by the Farmers to undertake the cultivation of these medicinal plants in the country. We in Sri Venkateswara College, The College with Potential for Excellence, University Of Delhi, Dhaula Kuan, New Delhi-110021 are engaged in the Mycorrhizal inoculation of Medicinal Plants in an effort to give the exact cultural practices to be used by the Farmers to get not only higher yields but also the quality medicinal plants for the Ayurvedic Pharmacies.

The Mother Nature has provided a wide range of the soil microbes for the benefit of the plants and all are active in the rhizoshere zone. Lord Brahma Jee, the Creator of the Universe, did not send the Man first when He created the Earth. Rather, He first sent the plants and to look after these plants, the soil microbes were created. There are a wide range of the soil microbes. Nature offers a wide scope to search and research for concealed, diversified and valuable organisms, which are important not only for the present but also to the future. There is a vast microbial flora inheriting the Earth and they are found in all types of soils, which are pure or cultivated, sands, desserts , thermal soils, snow covered soils and others. These microbes are responsible for the fertility of the soils and ever since the Green Revolution started in the mid sixties, with the excessive use of the chemical fertilizers coupled with the various agrochemicals without adding the required quantity of the organic manures, over the years, we have practically killed the microflora of the soils in the lands of Haryana and Panjab which are the" feeding bowl". The lands are practically barren. The micronutrients are in the precipitated form not available to the plants. The fertility of the soils is very low. The organic matter is very low. Till you do not add more quantity of the urea and DAP, you do not get the required yields. Infact, the yields of the crops per acre are declining in the subsequent years and the farmers of these states are fed up with the high costs involved in the cultivation of the wheat and paddy.

We have to go back to Nature Farming which was practiced here in the East since several Yugas in the Vedic Period and was so popular till recently based on the Rishi-Krishi Techniques and Vedic Krishi. The initiative has been taken by the group of retired and devoted scientists from ICRISAT in the form of a NGO called The Asian Agro History Foundation constituted by Prof. Y L Nene, former DDG, ICRISAT, Hyderabad(47, ICRISAT Colony, Brig.Sayeed Road, Secundrabad) who have translated the ancient scripts like Vrisha Ayurveda by Surpala, texts written by Rishi Parashar and Kashyap and now publishing a quarterly newsletter of the AAHF called The Asian Agro History which should go to each and every village and the Gram Panchayats and the Progressive Kisan Clubs so that the ancient wisdom of our country in the field of Agriculture is shared by all and we get rid of the problems of the multinationals companies coming very fast in the country and capture the market. These Newsletters are carrying articles which are the eye opener to all of us and it is the duty of the learned scientists of the Modern time to read these articles and take the results to the farmers who do not know about these ancient scripts. By adopting these time tested techniques only, we can achieve the sustainability in agriculture rather than popularizing the MNC products. These are various challenges which have to be adopted, demonstrated and let the farmers in the remotest area get benefited .

The dominating groups of the microorganisms are Bacteria, Actinomycetes, Fungi, Nematodes and Protozoas. But most of these organisms share a common character in being heterotrophic in their nutrition and there by depend on other organic , dead, or living organisms and inorganic source of nutrition for their survival and multiplication.

Apart from Bio-trophic and saprobic microbes, symbiotic microbes are also found in the soil. The microbial population of soil is , therefore, dependent on composite micro-ecological environments. Plant productivity , growth and yield are dependent on the beneficial association of microbes established around plant root system. Plant root-soil system is a very complex environment in which the effects of the plant on soil microorganisms and the effects of microorganisms on the plant are interacting interdependent, and highly complex. Plant root exudates released and break down products of the root and its environment feed the microbes and the microbes in turn often benefit the plant. Mycorrhizae are the composite structures formed by nutritional associations between fungal hyphae and roots of higher plants. The infected feeder roots are called mycorrhizae (myco=fungus;

rrhizae=roots). As this fungus plant association is ubiquitous, studies on plant root functions should also consider the effect of the mycorrhizal fungi on these functions. Fungi receive carbohydrates from the host plant while plants receive fastly mobilized phosphorus and other nutrients from fungus.

Seven types of mycorrhizal associations are known and mycorrhizal plants suggest saving of 50% of phosphorus if proper management is observed by deriving maximum mycorrhizal response.

Arbuscular mycorrhizal fungi are the naturally occurring fungal component of soil biota in most terrestrial ecosystems as well as in some aquatic systems (Stenland and Charvat, 1994). They are very common in annual plants but also occur in shrub and tree species. They represent an obligate symbiotic group distinct from the rest of soil microbial biomass. Therefore, they cannot complete their life cycle without the plant host. AM fungal symbiosis is a dynamic process and interaction affects all physiological aspects of the host. They are unique as they are partly inside and partly outside the root.

They may remain in the form of individual hyphae on the root surface or hyphal connections may be found between roots and substrates. However they penetrate into cortical cells of the root by secreting cellulolytic enzymes and grow there intracellularly and also intercellularly. They form special structures namely Vesicles (terminal swellings of hyphae), Arbuscules (dichotomously branched haustoria like structures) and normal hyphae. These are formed inside the root and they do not encounter competition and antagonism from soil microbes.

Soil factors and host genotypes are some of the parameters that affect biological interactions in the rhizosphere. The AM fungi occur abundantly in phosphorus and other nutrient deficient soils. 80%-90% of plants ranging from Bryophytes to flowering plants have the dynamic association of AM fungi.

Until recently AM fungi fall into the order Glomales in the Zygomycotina. But the taxonomy of this fungi continues to change as a result of new information derived from all scales (molecular, organismal, morphological and ecological).

The kingdom Mycota has been circumscribed by the use of morphological, biochemical and molecular studies. Schubler *et al*, (2001) based on the comprehensive SSU rRNa analysis, separated the AM fungi in a monophyletic clade, which is not related to any zygomycetous group, but probably shares a common ancestry with the Ascomycotina – Basidomycotina clade. They recognized a new fungal phylum based on natural relationship for the AM fungi, the Glomeromycotina with a single class Glomeromycetes (Cavalier-Smith, 1998). The class Glomeromycetes was circumscribed for the phylum containing more than 150 described species and some of which are synonyms (Walker and Vestberg, 1998; Walker and Trappe, 1993).

AM fungi cannot be cultured artificially due to their obligate symbiosis. Arbuscules are the key sites formed inside the cortical cells and are meant for nutrient exchange and remain active up to 15 days depending upon host's response. Vesicles are formed in/on inter and intracellular hyphae while spores/sporocarps and branched hyphae are formed outside the root and found associated with the soil. At present there are ten valid genera encountered namely *Glomus A, Glomus B, Paraglomus, Diversispora, Gigaspora, Scutellospora, Acaulospora, Entrophospora, Archaeospora* and *Geosiphon*.

Vesicles are not reported in *Gigaspora* and *Scutellospora* but extrametrical vesicles were reported. There are more than 150 AM fungal species reported colonizing 30, 000 receptive hosts. AM fungal taxonomy is more complicated and morphotaxonomic criteria viz, wall layers, ornamentation, subtending hyphae, shape, size, colour and other characters help to segregate genera and species. Presence or absence of the sporocarp is also an important criterion in identifying AM fungal genera.

Molecular and biochemical parameters such as cell wall composition, protein immunogenecity, Isozyme pattern, fatty and methyl ester profiles, direct sequencing of 18-SRNA gene and RAPD approach to detect polymorphism and other have been proved more useful. PCR-RFLP can be used for quantitative comparision. Morton and Redecker (2001) reported two new families of *Glomales, Archaeosporaceae* and *Paraglomaceae* with two new genera *Archaeospora* and *Paraglomus* based on concordant molecular and morphological characters.

AM fungi help in soil binding, efficient and effective mobilization of phosphorous and nutrients, restoration of disturbed land, mine soils, establishment of forest seedlings, commercial crops, economically viable plants, medicinal ornamental plants, aromatic plants, horticultural plants besides increasing N, P, K status in mycorrhizal plant and water uptake, improves plant defense against biotic and abiotic stress, increase yield and offer other benefits to host plants on inoculation with AM fungi.

Root exudation and soil factors are the important ecological phenomenon, which can manipulate plant and microbial succession. Mycorrhizal plant exudates include many compounds useful to host plant including growth-promoting substances. Various techniques used for the isolation of AM fungal spores from the soil such as floating technique, decanting and sieving, gradient centrifugation, monoclonal and polyclonal methods are employed. Viability of AM fungi is of great significance in AM fungal application.

Colonization of host plants, crop plants by AM fungi and its degree of colonization form an important factor in mycorrhizal dependency studies.

Application of AM fungi as biofertilizer has been limited by the difficulties of obtaining large quantities of pure inoculum of the fungal symbiont. Resting spores are extracted from soil and are often used as inoculum. To grow AM fungi through nutrient film technique, aeroponics, root organ culture, Ti/Ri-plasmid and other techniques resulted in limited success. However, soilless medium seems to avoid contaminants hence soil-pot culture with a suitable host has been the most practicable method employed for multiplication of AM fungi. AM fungi, which are the beneficial fungi, need to be conserved through various scientific methods. The word 'mycorrhiza' denotes a symbiotic, non-pathogenic or weakly pathogenic association of a fungus and the roots of a plant (Kirk *et al*, 2001). Mycorrhizae are the structures formed by the association of plant roots with soil fungi. Such root fungus associations were found to be normal on the root system. The mycorrhizal fungi receive carbohydrates from host plant, while host plants receive fastly mobilized phosphorus and other nutrients through mycorrhizal roots.

Frank (1885), a German botanist who first noted the existence of a characteristic, beneficial association in the roots of *Cupuliferae* coined the term mycorrhiza. Since then our knowledge on mycorrhizae and plant relationship has grown and lots of valuable information has come to forefront. Word wide, a lot of research work has been done to understand the functioning of Mycorrhiza. There are more than 40, 000 research papers which have been published in the various scientific journals but unfortunately, the results have not gone to the farmers. The scenario is changing now with technologies being transferred to the private sector.

Mycorrhizae were prevalent through out the world and they are omnipresent. Based on location, seven types of mycorrhizal association are known namely-*Ecto-mycorrhizae, endo* or *arbuscular mycorrhizae, ecto-endo mycorrhizae, arbutoid mycorrhizae, monotropoid, ericoid* and *orchidoid mycorrhizae*.

In *ecto-mycorrhizae*, morphology of the infected root system is changed. The invaded fungus forms a mantle of variable thickness, colour and texture surrounding the rootlets. Root hairs usually do not develop instead in the infected roots, hyphae radiate from the mantle into the soil. As a result of this invasion, the infected root system becomes stunted in growth, swollen, often forked or even coralloid through repeated dichotomous branching. The affected roots sometimes become brightly coloured depending on the color of the fungal symbionts, secreting pectolytic enzymes, the fungi penetrate between epidermis and cortical cells of the host root and grow there intercellularly calculably forming a network of hyphae which is then called "hartig's net".

The important ecto-trophic mycorrhizas are common in forest trees including genera such as: *Eucalyptus, Populus, Salix alba, Cedrus, Deodar, Picea sp., Pinus, Larix, Tsuga, Fagus, Quercus, Shorea* and others. There are over 1000 species of fungi involved in forming ecto-mycorrhizae as about 2000 species of woody plants.

Ectomycorrhizae occur naturally as feeder roots on most conifers including all species of the family. *Pinaceae* as well as some angiosperms such as members of the *Fagaceae, Betulaceae, Juglanaceae* and *Myrtaceae*.

Endo-mycorrhizae

In contrast to ecto-mycorrhizae, the morphology of the infected root in endotrophic mycorrhizae generally remains unchanged or it may become beaded, but external sheath of fungal hyphae is absent and root hairs develop normally. Vesicles, arbuscules and inter/intracellular hyphae develop in the root cortex. Spores and sporocarps are formed on mycelium present outside the root.

Ecto-endo-mycorrhizae

In this type both the ectotrophic and endotrophic are present as observed on roots of pine and beech where, infection is typical ectotrophic but with intracellular penetration of hyphae.

Arbutoid mycorrhizae

These *mycorrhizae* were seen in the roots of a tree called *Arbutus*.

Montropoid mycorrhizae

These *mycorrhizae* were associated with the roots of a genus *Monotropa*.

Ericoid mycorrhizae

These are different type of *mycorrhizae* particularly seen in the members of *Ericaceae*.

Orchidoid mycorrhizae

These *mycorrhizae* were seen in the roots of Orchids. They are characterized by the presence of *glomerules*. That is why they are also known as *Glomerate mycorrhizae.*

Description of different tpes of mycorrhizae as classified by Harley (1989) has been outlined in table 1.

Table 1 : Characteristics of important kinds of *mycorrhizas* (Harley, 1989)

Character	Kinds of Mycorrhizae						
	vesi-cular	Ecto-myco	Cto-endo-myco	Arb-utoid-myco	Mono-tropoid-myco	Eri-coid-Myco	Orchid-oid-myco
Fungus							
Septate		+	+	+	+	+	+
Aseptate	+	+	-	-	-	-	+
Hyphae							
Enter cells	+	-	+	+	+	+	-
Fungal Sheath							
Present	-	+	+or-	+	+	-	+
Hartig's net formed	-	+	+	+	+	-	-
Hyphal coils in cells	+	-	+	+		+	
Haustoria							
Dichotomous	+	-	-	-		-	-
Non Dichotomous	-	-	-	-	+	-	+or-

Among the seven types of *mycorrhizae Arbuscular mycorrhizae* are the most prevalent ones. They are omnipresent and occur in all groups of plants eg: Bryophytes (esp. *Heapaticae*), Pteridophytes, Gymnosperms (except *Pinaceae*) and Angiosperms and are found in most ecosystems including Dense rain forests, open woodlands, Savannas, Grass lands, Heaths, Sand dunes and Semi-arid deserts.

AM fungi form a very unusual or exceptional structures namely arbuscules, vesicles and hyphae. They penetrate into cortical cells of the root by secreting cellulolytic enzymes and grow there intracellularly and also intercellularly.

Arbuscules are finely branched, tree like intracellular haustoria like ephemeral structures and vesicles are inter/intra cellular, globose, cylindrical or lobulate lipid filled storage structures. Mosse (1953) was the first scientist to demonstrate experimentally that AM fungi can produce arbuscules and vesicles. That is why they were initially named as vesicular *arbuscular mycorrhizal* fungi or VAM fungi (Smith, 1995). However all the genera do not produce vesicles but arbuscules are produced by all of them, therefore they are now called as *arbuscular mycorrhizal* fungi or AM fungi (Walker, 1995). The hyphae are aseptate and may remain on the root surface or hyphal connections may be found between roots and substrates.

AM fungi are wide spread in occurrence and occur in all types of soils, but abundantly found in soils deficient in phosphorus and other nutrients.

The majority of higher green plants and large number of fungi are involved in *mychorrhiza* formation . About 80% species of Angiosperms, 100% of the Gymnosperms and 70% of the Pteridophytes were found to be *mycorrhizal* (Harley and Jarley, 1987). AM fungi do not grow in axenic culture. They must be associated with living plant roots. Some host plants with fibrous root system are highly susceptible for AM fungi eg. *Allium cepa, Arachis hypogea, Cenchrus, Eleusine coracana, Sorghum vulgare* and other plants. Such plants are used all over the world for the multiplication of the AM fungi.

There are several consistent structural/functional relationships associated with all arbuscular mycorrhizal symbiosis, which are critical to understand their role in ecosystem dynamics.

The arbuscule is an important point of contact for the exchange of resources between plant and fungus. Both organisms retain a membrane to separate their adjacent cells (Cox and Tinker, 1976). Enzymes such as phosphatases are concentrated along the interface, presumably facilitating nutrient movement from fungus to plant.

Intraradical structures, because of their existence create an altered physiology of the host. Several research laboratories are presently studying the mechanisms whereby signals are exchanged between host and fungus, which are regulating their development (Giovannetti *et al*, 1994).

Associated with the hyphae that penetrates a root is an external matrix that radiates out into the soil. There are important architectural features associated with these extra metrical hyphae. These include runner or arterial hyphae that are thick walled and often traverse long distances to invade uncolonized roots (Mosse, 1959; Friese and Allen, 1991).

These hyphae may be a major carbon sink in storing and transporting a large amount of energy necessary for colonizing new roots. These hyphae may also serve as bridges between plants, transporting resources between individuals (Read, 1992).

When a hypha penetrates the root, the runner hypha produces finer branched hyphae that radiate out into the soil matrix. Friese and Allen (1991) suggested that this pattern was a distinct architectural feature of the symbiosis that is critical to *arbuscular mycorrhizal* functioning.

Vesicles develop to accumulate storage products in many AM fungal association as they are intiated soon after the first arbuscules, but continue to develop when the arbuscules senesce.

Vesicles are hyphal swellings in the root cortex that contain lipids and cytoplasm and these may be inter or intra cellular. Vesicles can develop thick walls in older roots and may function as propagules (Biermann and Linderman, 1983). Some fungi produce vesicles which are similar in structures to the spores that are produced in the soil, but in other cases they are different.

TYPES OF ARBUSCULAR MYCORRHIZAL FUNGI

There are two types of *arbuscular mycorrhizal* fungi (Barker *et al*, 1998).

1. Arum type
2. Paris type

Arum Type

Many herbaceous plants exhibit the Arum colonization type, which involves extensive intercellular growth of the fungus as it penetrates the root cortex, followed later in the colonization by formation of arbuscules.

Paris Type

In Paris type of colonization growth into the root is slow being primarily intracellular and the fungus forms coils inside each cell with rate or minimally structured arbuscules (Gallaud, 1905).

The term vesicular arbuscular mycorrhiza (VAM) was originally applied to symbiotic associations formed by all fungi in the *Glomales*, but because a major suborder lacks the ability to form vesicles in roots. *Arbuscular mycorrhiza* (AM) is now the preferred acronym. The order *Glomales* is further divided into families and genera according to the method of spore formation. The spores of AMF are very distinctive (Morton and Benny, 1990). Schubler *et al*, (2001) based on comprehensive SSU rRNA analysis and on the basis of natural relationship of *Arbuscular mycorrhizal* fungi and the related fungi recognized a new fungal phylum – *Glomeromycota*.

SIGNIFICANCE OF *ARBUSCULAR MYCORRHIZAL* FUNGI

AM fungi play an important role in the improvement of plant growth. Most vascular plants require *mycorrhizae* for their survival. They are vital for the uptake and accumulation of ions from the soil and their translocation to the hosts because of their high metabolic rate and strategically diffused distribution in the upper soil layers. They produce growth promoting substances, help in drowught tolerance and impede soil-borne/root pathogens. The fungus serves as a highly efficient extension of the host root system (Bolan, 1991). Minerals like N, P, K, Ca, S, Zn, Cu and Sr are absorbed from the soil by AMF and are translocated to the host plant (Mosse, 1957).

Improved nutrient uptake especially phosphorus (P) is the primary cause for improved plant productivity. Earlier studies (Baylis, 1967; Sanders and Tinker, 1971 and Marchener and Dell, 1994; Fidelibus *et al*, 2001) have revealed that mycorrhizal roots absorbed phosphate from low P status at a greater rate per unit length of root than non-*mycorrhizal* roots. The high efficiency in nutrient uptake by *mycorrhizal* roots is mainly due to the activity of the hyphal network developing from the roots into the surrounding soil thus increasing the absorptive root surface area for nutrients and their translocation to the host plant (Allen *et al*, 1981; Berta *et al*, 1995).

AMF have great potential to enhance plant growth by increasing nutrient uptake (Bagyaraj, 1992; Vasantha Krishna and Bagyaraj, 1993). They also act as potential factors in determining diversity in ecosystem (Giovannetti and Gianinazzi – Pearson, 1994).

AM fungi are conditioned by the edaphic factors. It has been found that pH (Daniels and Trappe, 1980) plays an important role in the distribution of AM fungi normally it has been observed that acidic to neutral soils have a greater role to play. Soil moisture exerts influence on mycorrhizal associations (Redhead, 1975. Species and strains of AMF differ in their ranges to soil, physical and chemical properties (Abbott and Robson, 1991; Mukerji, 1995).

Am fungal association has been reported in field crops (Bagyaraj *et al*, 1979) cereal crops (Ammani *et al*, 1985; Gerdemann, 1961; Hetrick and Bloom 1983; Jensen and Jacobson, 1980) Legumes (Krishna and Bagyaraj, 1982a; Parvathi *et al*, 1984; Rao and Parvathi (1982), Vegetable crops (Creighton Miller *et al*, 1986; Daft and El-Giahmi, 1975; Manjunath and Bagyaraj, 1984), oil seed plants (RAo and Parvathi, 1982; Vijaya Lakshmi and Rao, 1987, 1988; Manoharachary and sulochana, 1989) herbaceous plants (Ardey and Mukerji, 1986), mangrove plants (Mohan kumar and Mahadevan, 1986), and xerophytic plants (Singh and Varma, 1981).

With increasing awareness about the beneficial aspects of herbal medicines, there is a surge in the demand for medicinal plants. The role of AM fungi in enhancing plant growth as well as improving host resistance to certain plant diseases is well documented (Wright and Milliner, 1994). The incidence of mycorrhizal associations in medicinal plants is less documented compared to studies on forestry and crop species (Bagyaraj, 1995; Sharat Babu and Manoharachary, 2003).

WHY *MYCORRHIZA* UNIQUE OVER OTHER BIOFERTILIZERS?

Mycorrhiza is the only known fungal system which is categorized as a biofertilizer. Its hypae can extend much beyond a few meters away the depletion zone and thus can acquire nutrients from a much wider soil area. In soil, *mycorrhiza* produces vegetative structures like chlamydospores and zygospores which become dominant during the period of environmental stress and germinate with the return of favourable conditions. Thus, they are better equipped for combating the unfavorable conditions and have larger shelf lives as compared to bacterial systems.

- *Mycorrhiza* is broard spectrum nonspecific organism. A single species is known to colonize 85 % of the land plants.
- It has broard ecological adoptability and is known to occur in desserts as well as arctic, temperate, tropical and other inhospitable habitats.
- It offers upto 50 % reduction in the phosphatic fertilizers application and under organic conditions where organic manure is applied and the only source for N, P, K, will dissolve the combined form of the organic Phosphorous.
- It facilites better uptake of nutrients like P and immobile trace elements like Zinc, Cobalt, , Manganese, .Iron, Copper. Molybdenum etc leading to better nutriention for the plants.
- It offers to tolerance against a range of soil stresses like heavy metals toxicity, salinity, drought and high soil temperature. This enhances the chances of plant survival immensely.
- It offers higher resistance to various soil and root born pathogens thus becoming a potential disease control agent.
- It increases the rate of photosynthesis and hence improves plant growth.
- The AM Biofertilizer is known to increase the nitrogen fixing potential of legumes when given together with *Rhizobium*. The *Mycorrhiza* first stimulates the nodules bacteria in a sequential process by increasing the tissue

phosphorous contact, this results in improved nodulation. There are also reports of positive interaction between *Azotobacter* and AM fungi. AM colonization favorably affects the population of these free living N- fixing bacteria and thus stimulates better growth of the plants. But these dual inoculation effects have to be seen in the medicinal plants.

Abdul – Khaliq and Janardhan (1997) reported AM fungal influence on the productivity of cultivated mints. About- Nasar (1996) has reported the effects of AMF on *Tagetus erecta* and *Zinnia elegans,* while Manoharachary and Ranaganayaki (2003) have studied improvement in the medicinal plants through AM fungi. Gupta *et al* (2000) worked on AM fungi in association with *Ocimum* Spp. Harris and Clapperton (1997) have studied the influence of AM fungi on the success of weed biological control with insects. Kothari and Singh (1996) have studied the response of *Citronella Java* to AM fungi and soil compaction in relation to phosphorous supply. Marks (1991) studied casual morphology and evolution of *mycorrhizas*. Morton (1988) described the taxonomy of AM fungi in detail. Payal-Mago *et al* (1994) reported AM fungal associations in *Lamiaceae*. Pfleger and Linderman (1994) studied the relationship between mycorrhizae and plant health. Rao *et al* (1989) studied VAM in medicinal plants. Rao *et al* (1996) reported the effect of the integration of *Calotropis procera* leaf and *Glomus fasciculatum* on the management of *Meloidogyne incognita* infesting Tomato. Tanushree *et al* (1994) reported the application of AM to increase the alkaloid contents in *Catharanthus roseus*. Tanvir *et al* (1994) worked on AM associations in some cultivated medicinal plants. Untch and Weber (1995) reported the structure of the AM fungi in *Ceropegia dichotoma, Ceropegia fusca* and *Periploca laevigata*. Weber *et al* (1995) studied the anatomical structures of the AM in the *Apocynaceae*.

Ecology of AM fungi has been extensively studied by various researches in different areas. Akhilesh Kumar *et al* (1999) reported AM and revegetation of coal mine spoils and Chang and Young (1992) have studied the ecoogy of AM fungi in tea garden soils. Johnson (1993) discussed in detail about the relation between fertilization of soil and mycorrhizae. Galvez *et al* 2001) reported the effect of tillage and farming system upon AM fungal populations and nutrient uptake of maize.

Gupta *et al* (2002) reported the effect of the AM fungus Glomus fasciculatum on the essential oil yield related characters and nutrient acquisition in the crops of different cultivation of menthol mint (*Mentha arvensis*) under field conditions. Gupta *et al* (2000); Manoharachary and Ranganayaki (2003) made significant research and development with reference to biofertilizers in Chhattisgarh and Andhra Pradesh respectively, Meena *et al* (1991) noticed the natural incidence of AM in hot pepper. Pacioni and Puppi (1988) studied the ecology of vesicular arbuscular endophyes in Central Italy. Srinivas *et al* (19990 reported the dynamics of AM fungi of Godavari beld forest. Varsha *et al* (2001) surveyed the occurrence and distribution of AM fungi in coastal sand dune vegetation of Goa. Wani and Konde (1996) reported the vesicular arbuscular mycorrhizal association in garlic while Kunwar and Manoharachary (1999); Kunwar *et al* (1999) studied AM fungi in Ginger and Garlic. Schaffer and Peterson (1993) reported modifications to clearing methods used in comhination with vital staining of roots colonized with vesicular arbuscular mycorrhizal fungi. Schenck and Perez (1990) introduced the isolation and culture of vesicular arbuscular mycorrhizal fungi. Schenck and Perez (1990) introduced the isolation and culture of vesicular arbuscular mycorrizal fungi.

Occurrence

AM are the most common type of *mycorrhizae* involving majority of plants. AM Fungi are known to occur globally in a broad range of dissimilar environments (Gerdemann and Trappe, 1974; Koske, 1987) from the arctic to the tropics and occupy a wide range of ecological nitches (Srivastava *et al*, 1996)

Distribution

AMF organisms are ubiquitous in distribution and occur in all kinds of ecosystems ranging from Arid dry lands, Grass lands, Tropics etc. *Arbuscular mycorrhizas* can also be found in unexpected plant groups and locations, viz arboreal habitats (Janos, 1993); aquatic trees (Khan, 1993); parasitic plants (Stenlund and Charvat, 1994); Proteaceae (Bellgard *et al*, 1994) and in extreme habitats like deserts (Jacobson *et al*, 1993 and the Anaconda (Dhillion *et al*, 1995). Importantly Arbuscular mycorrhiza appear to be missing islands formed in arctic soils (Christie and Nicolson, 1983) while DeMars and Boerner (1995) found that the plants without mycelium in the Antarctic readily formed Arbuscular mycorrhiza in pot cultures, where inoculum was present.

Dispersal

AM fungi are readily dispersed by rodents (Maser *et al*, 1978; McGee and Baczocha, 1994) but poorly by wind in humid conditions (Allen, 1988) which predominate in the arctic or on islands at high altitudes. In arid lands where soils are dry, reinvasion is relatively rapid (Allen, 1988). In a cold desert steppe, Arbuscular mycorrhiza rapidly invaded a disturbed site, primarily by wind (Alen, 1988).

FACTORS AFFECTING AM FUNGI

Edaphic factors can affect AM fungal spore germination, colonization of host roots and ability of the AMF to influence the growth and physiology of the host plant. A high degree of edaphic specificity in indigenious assemblages of AM fungi form contiguous soil has been observed. Soil physico chemical characteristics viz, soil moisture, soil temperature, soil pH, soil texture, nutrient levels alter the occurrence, distribution and effectiveness of *Arbuscular mycorrhizae*.

SOIL-PHYSICAL CHARACTERISTICS

Soil Moisture

Soil moisture influences *Arbuscular mycorrhizal* colonization and growth (Redhead, 1975). Both excessive and low soil water potential decreased *mycorrhizal* colonization and sporulation (Nelson and Safir, 1982a). Studies by Anderson *et al* (1983); Allen and Allen (1984); Douds and Schenck (1991); Muthu Kumar *et al* (1994; 1998) also suggested the significant effect of soil moisture on root colonization, Arbuscular mucorrhizal fungal spore germination and spore abundance.

Sing and Tyagi (1989) reported the root colonization of *Aeschynomene indica* by indigenous AM fungi grown under normal moisture conditions. On the other hand no AM fungal colonization occurred in this plant uprooted from water logged conditions. However, when such plants were transplanted to potted soil conditions, after scissoring lateral roots and provided with normal moisture conditions. Thereafter the new roots developed were highly colonized by AM fungi. These results have indicated the requirement of normal conditions for better Am fungal root colonization which is reduced due to higher moisture level/flooding. Rice cultivars remained colonized with AM fungi in their roots during their growth period under high moisture level and water logging conditions (Gupta, 1996); Singh, 1996).

Soil Temperature

Soil temperature alters the physiology of mycorrhizal symbiosis by influencing the root morphology, nutrition and growth. Studies on effect of root zone temperature on VA mycorrhiza formation have revealed that maximum formation of VA mycorrhizae occurred at 18.5 –24 C soil temperature. Mycorrhiza formation was moderate at 7.5 C but reduced/ inhibited at or above 29.5 (Parke *et al*, 1983). High soil temperature favours colonization and sporulation (Furlan and Fortin, 1973) while lower temperature favours arbuscule formation (Schenck and Schroder, 1974). Low AM fungal spore germination have been reported in temperate soils where temperature is low as compared to tropical soils. *Glomus* and *Acaulospora* spores germinate between 20-25 C and of *Gigaspora* around 30 C (Daniels and Trappe, 1980). Some species of *Glomus* are shown to be more adapted to high temperature that others (Schenck and Schroder, 1974; Schenck *et al*, 1975). Koske (1981) found optimum temperature for germination of *Gigaspora gigantean* from Rhode Island to be 30 C.

Soil pH

The distribution of AMF has been shown to be correlated with soil pH (Abbott and Robson, 1977; Wang *et al*, 1985). Varying soil pH affects the development and functioning of *Arbuscular mycorrhizae* (Mosse, 1972; Abbot and Robson, 1985; Hayman and Taveres, 1985) and also the germination and hyphal growth of some AMF (Green *et al*, 1976; Hepper, 1984). Normally it has been observed that acidic to neutral soils have a greater role to play, because they favour the germination and multiplication of AMF (Furlan and Fortin, 1973).

Some *Glomus* species are very common in neutral or alkaline soils but few in acidic soils whereas species of *Acaulospora* are usually found in acidic soils. Lower soil pH may have positive, indifferent or negative responses on *Mycorrhiza* formation. Soil pH over a range of 4.8-8.0 significantly influenced germination of spores of *Glomus epigeum* (Daniels and Trappe, 1980)

Mycorrhizal association potential decrease with increase in initial soil pH (Al-Agely and Reeves, 1995). This is in conformity with (Powell and Bagyaraj, 1984, 1984) that AM fungal spore germination occurs within a range of pH that is acceptable to plant growth. *Mycorrhiza* formation depends both on soil pH and moisture content. However no relationship was found between mycorrhizal colonization and soil pH and moisture by some others (Peat and fitter, 1993; Wang *et al*, 1985). Though *Arbuscular mycorrhiza* formation was more at pH6 than ecto or ericoid *mycorrhizae*. AM colonization slightly decreased with decreasing soil pH though they are tolerant to broad soil pH (Hayman and Taveres, 1985; Hiejne *et al*, 1996, Wang *et al*, 1985). Optimum pH for colonization of *Glomus fasciculatum* in maize root was found to be between pH 5.6-6.2 and in wheat root between pH 6.7-6.9. Read *et al* (1976) have also found high percentage of AM colonization in *Festuca* ovina irrespective of differences in soil pH.

Danielson and Visser (1989) could not demonstrate a correlation between length of root colonized by AM and soil pH. The type of structures formed by AM also differs at different pH (Duckmanton and Widden, 1994).

The relationshop between soil pH and mycorrhization is complex and depends on the plant species and also soil type, forms of phosphorus and fungal species involved (Walker and Smith, 1984).

Soil Texture

Texture is an important characteristic feature of soil affecting drainage conditions, water holding capacity, amount and size of the pores and plant root development. The basic structure of soils is a composition of mineral matter, organic matter, water and air. The mineral matter is composed of inorganic particles varying in size from stone and gravel to powder. These inorganic particles, separated according to size are referred to as soil separates. The U.S. Department of Agriculture (USDA) recognizes three major groups of soil separates: sand (2.0 – 0.05 mm), silt (0.05 – 0.002 mm) and clay (<0.002mm). Generally red sandy soils harbour more number of AM fungal spores as they are well aerated and clayee soils with less aeration and more water holding capacity cannot harbour more number of AM fungal spores.

CHEMICAL CHARACTERISTICS: (SOIL NUTRIENT STATUS)

Phosphorus

Phosphorus is generally considered to be the most important factor which limits plant growth in most tropical soils, which could be supplied by mycorrhizal association because many abiotic and biotic factors can restrict its mobility and availability in soils (Harley and Smith, 1983; Hayman, 1983; Marschner, 1986 and Bolan, 1991). Reduction in AM fungal benefit occur with increasing soil phosphorus (P) levels (Schweiger *et al*, 1995). Several studies have suggested that root colonization by AMF is inhibited at high P level, because of decreased root exudation (Ratnayake *et al*, 1978; Graham *et al*, 1981; Muttu Kumar *et al*, 1994 and Udaiyan *et al*, 1996). Mycorrhizal associations can enhance the uptake of "P" by plant roots particularly in low fertility soils (Tinker, 1978; Bolan, 1991). AM fungal hyphae are able to absorb "P" directly from the soluble "P" pool in the soil and translocate it to the host roots (Cooper and Tinker, 1978 and 1981; Rhodes and Gerdemann, 1980).

Mycorrhizal plant can actively hydrolyze pyrate, the major source of organic phosphorus than the non-mycorrhizal plants both in soil (Gianinazzi *et al*, 1981) and in auxeinic culture (Allen *et al*, 1981).

Depending on soil "P" status, mycorrhizal infection may increase or decrease total and specific root length in different plant species (Gnekow and Marschner, 1984; Price *et al*, 1989).

The effect of soil "P" and relative benefit of mycorrhizal infection has been extensively studied (Bolan, 1991, Bowen, 1980; Bryla and Duniway, 1997; Harely, 1989; Hayman, 1982; Deucy and Paul, 1983; Marschner and Dell, 1994; Safir and Duniway, 1982.

The reports suggest that the increased levels of "P" availability in soils show a negative effect on the mycorrhizal population, development and symbiotic establishment. Infected roots and high spore numbers are found most commonly in soils of low – moderate

"P" content whereas high in "P" have been found to contain few AM fungal spores and infected roots (Azcon *et al*, 1978).

Nitrogen

AM fungal plants can derive nitrogen from both organic and inorganic sources that are not available to non-mycorrhizal plants (Ames *et al*, 1984; Barea *et al*, 1989). Uptake of N15 from labeled NH4 satls have been reported by Johnson *et al* (1992). Several workers from different parts of the world have reported increased N2 fixation and plant growth by Arbuscular mycorrhizae (Daft and EL-Giahmi, 1974; Islam *et al*, 1980; Manjunath and Bahyaraj, 1984; 1986). There is no direct evidence that AM can fix atmospheric N2, but increased growth and increased concentrations of N2 in the Plant tissue have been recorded (Barea *et al*, 1980).

Potassium

Generally low levels of Potassium (K) in soils support high AM population which is in accordance with the earlier observation (Harley, 1980). Recently similar kind of results were shown by Sharath Babu (2002) in *Tylophora indica*.

Other Nutrients

Both soil "P" levels and mycorrhizal colonization can influence uptake of nutrients such as Calcium (Ca), Magnesium (Mg), Ferrous (Fe), Copper (Cu), Molybdenum (Mn) and Zinc (Zn). *Mycorrhizal* colonization cured Zn efficiency in Peach seedlings and Apple (Gilmore, 1971; Benson and Covey, 1976). Improved Zn uptake has been reported in maple (Daft and Haeskaylo, 1977), Potato (Swaminathan and Verma, 1979), Metcalf (1978) reported a strong correlation between *mycorrhizal* infection and ability of cotton plants in the uptake of Zn. Apart from Zn even Cu, Fe, and Mn absorption from soil is increased by *mycorrhizal* plants. *Mycorrhizal* plants always contained higher quantities of mineral elements like Zn, Cu, Fe and Mn than non-mycorrhizal plants (Krishna and Bahyraj, 1984; Saif, 1987).

Am fungal ecology is an important aspect in order to study biodiversity, physico-chemical factors environmental variables and host plants influence the occurrence, seasonal studies were not extended to *Apocynaceae* and *Asclepiadaceae* host plants in relation to AM fungal association. Therefore an attempt is made.

Arbuscular myucorrhizae are cosmopolitan in distribution in soils supporting crop plants, forest vegetation, Bryophytes, Pteridophytes, Aquatic plants and other plant groups.

So far approximately 164 species AM fungi are recognized and many more are awaiting their discovery (Anonymous, 1997; Mukerji, 1996; Schenck and Perez, 1990). There are about 115 species of AM fungi that have been reported from India (Mukerji *et al*, 2002).

Beneficial aspects of AMF—The benefits or advantages of AMF can be differentiated into two categories.

1. Nutrient benefits
2. Non-nutrient benefits

Nutrient benefits

a) AM fungal association can enhance the uptake of Phosphorus by plant roots particularly in low fertility soils (Tinker, 1978; Bolan, 1991).

b) AM fungal plants can derive Nitrogen from both organic as well as inorganic sources that are not available to non – mycorrhizal plants (Ames *et al*, 1984; Barea *et al*, 1989).

c) They stimulate the uptake of minerals like N, K, Ca, S, Zn, Cu, Fe and Sr from the soil and translocate them to the host plant (Mosse, 1957). AMF have great potential to enhance plant growth by increasing nutrient uptake (Bagyaraj, 1992; Vasantha Krishna and Bagyaraj, 1993).

d) They also act as potential factors in determining diversity in ecosystem.

Non-nutrient Benefits

a) AMF improve the soil physical characteristics. They can bind and aggregate soil particles through the extensively growing mycelium (Sutton and Sheppard, 1976). Further the mycelium of AMF not only loosely aggregates soil particles but also the hyphae bind to them through amorphous polysaccharides. The mycelium also contributes to the formation of water-stable aggregates necessary for good soil tilth (Jeffries and Barea, 2000).

b) AM fungi are known to improve drought tolerance in plants (Aldon, 1975; Lindsey *et al*, 1977; Nelson and Safir, 1982a).

c) The mycorrhizal colonization affects the plant root in many ways. For example disease resistance and root longevity may be changed. Mycorrhizal plants are known to be less susceptible to wilting and transplant shock, than non-mycorrhizal plants (barrows and Roncadori, 1977).

d) Menge *et al*, (1978); Levy and Krikum, (1980); Janos, (1980); Hardie and Leyton, (1981); Sieverding, (1981); Biermann and Linderman, (1983); Cooper, (1984); Michelson and Rosendahl, 1990) and Allen and Allen, (1980) have reported increased cytokinin activity in leaves and roots.

e) AMF protect the plants from pests (nematodes) and fungal root pathogens (Azcon-Aquilar and Barea, 1996) they counter act the nematode infections by depressing root penetration and larval development (Sikara, 1978).

f) AMF plants show greater tolerance to heavy metal toxicity, high soil temperature; adverse soil pH, high acidity etc. (Shetty *et al*, 1995; Weissenhorn and Leyval, 1995; Pawlowska *et al*, 1996; Chaudhry *et al*, 1998; Chaudhry *et al*, 1999).

g) AMF produce enzymes, vitamins, cytokinins and other compounds that increase rootlet size and longevity.

h) The fungus serves as a high efficient extension of the host root system (Bolan 1991).

i) These fungi were determined to reduce plant stress such as that caused by root- infecting microorganisms, saline soils, and lack of soil moisture, high soil temperature and other factors.

j) AMF can function as biofertilizers, bioregulators and bioprotectors. The AMF are obligate biotrophs that play an important role in natural and agroecosystems by enhancing plant growth (Hayman and Taveres, 1985).

Arbuscular mycorrhizae are widespread in soils and often growth of Arbuscular mycorrhizal plants is increased in comparision to non-*mycorrhizal* plants, particularly at low availability of phosphorus (P). This beneficial effect on plant growth has largely been attributed to higher uptake of P (Rhodes and Gerdemann, 1975), Zn (Swaminathan and Verma, 1979) and Cu (Pacovsky, 1986).

AM fungal colonization is the index of *mycorrhiza* formation. These fungi forma symbiotic relationship with plant roots, and the fungal symbiont becomes a major interface or connection between soil and plants, therefore studies on mycorrhizal systems need to consider soil as well as plants and fungi. Edaphic factors can affect sport-germination, colonization of host roots and availability of the AM fungi to influence the growth and physiology of the host plant.

Recently Sambandanak K of Avviyer Govt College of Women, Karaikal, Pondicherry have carried carried out a very interesting study on the effect of Tsunami on the plants in the coastal area of Karaikal, Pondicheery. Due to high rising tidal waves contaminated with high concentration of the salts accumulation in the Karaikal coastal area, first all the local vegetation was destroyed. A practically high sress condition was created because of the excessive salt accumulation but Nature is so vital and does wonderful things for the survival of the plants. After Tsunami , the root infection by the local mycorrhizal spores was found to be much higher than before Tsunami and this resulted in the regeneration of the plants found locally in that area along the coastal area affected by the Tsunami. Thus under the stress conditions of high salinity, mycorrhiza inoculation is enhanced and the plant wealth is saved by the Nature.

As the spores of the *Mycorrhiza* multiply only inside the hairy live roots of the plants, this fungus unlike many other fungi can't be cultured on the synthetic medium for large scale multiplication and this has been a bib hurdle world wide that pure inoculum of the various mycorrhizal spores could not be provided to the farming community by the private sector. Scientists are producing the inoculum for their own research work by the pot culture technique which involves the inoculation of the host plants where the % root infection by the mycorrhiza spores is very high with the mycorrhizal spores to be multiplied in the sterilized potting mixture, allowing the plants to grow for three months and the roots are harvested , dried , cut into small pieces and the roots will have the mycorrhizal spores and these roots will become the Mycorrhizal inoculum for further inoculation studies. But by this method, one will get lesser no. of the spores density and chances for the infection are much more. However, in our own country, the scientists of *Mycorrhiza* Research Centre, The Energy and Resource Institute, New Delhi under the guidance of the young devoted and commited scientist, Dr. Alok Adholeya, have developed a Swedeshi Technology for the mass scale production of the Mycorrhiza Inoculum much more rapidly, in large spore density and sterile condition by carrying out reseach over a period of 20 years. Dr Alok was also awarded by the President of India, Dr. A.P.J. Abdul Kalam, in 2005 for developing the new technology finding extensive use in promoting not only the organic farming but saving the foreign exchange for the production of the phosphatic fertilizers as the use of the *Mycorrhzia* saves the consumption of the phosphatic fertilizers upto 40% and still producing the yields which are higher by 15-40 % than the control. The technique is called as Root Organ Culture Technique which they have already transferred to four Institutes in the private sector covering all the four parts of the country and the total production of the *Mycorrhizal* inoculum has gone from 70 tons per year in 1995 to 4000 tons a year in 2006. More and more industrial houses should come forward to take the technology so that *Mycorrhiza* is made available to each and every farmer even in the remote villages. The young microbiologists should make the consortium and be involved in the production of the this biofertilizer.

Research carried out on the Mycorrhizal inoculation of Medicinal Plants in Sri Venkateswara College, (University of Delhi), The College of Potential with Excellence.

A research project" Biodiversity of AMF in the Medicinal Plants found In Wild Habitation in the Reserved Saraswati Forest Distt. Kaithal and Aravali Ranges, Distt.Gurgaon, Inoculation of the Medicinal Plants and Comparative Phytochemical Studies" was SANCTIONED TO Sri Venkateswara College, University of Delhi in April 2002 with Dr. Yudhvir K.Bhoon As Principal Investigator which was completed in October 2005 and we got some interesting results which are being discussed below.

1. *Silybum marianum*

The experiment was conducted in a view of the extensive use of *Silybum marianum* in Allopathic system for the treatment of Hepatitis B, C which are virus based and is being adopted by Indian System of Medicine also and also because the fact that the plant is being cultivated commercially in India to meet the international market demand. The growth of the plants raised from direct sown seeds which were inoculated with *Ecorrhiza* at the time of sowing of the seeds was much satisfactory than non inoculated plants. The total height of the plants, leaf area, the branching, the collar girth of the plants was higher in inoculated plants. The flowering was observed 10 days earlier in the inoculated plants. The colour of the seeds was dark black in the inoculated plants while in the control, it was light brown and variegated. The seeds of the inoculated and control plants were subjected to HPLC Analysis for the Silymarin contents and the seeds from the inoculated plants were found to have silymarin contents1.88 % while the control seeds were having 1.77 %. There is a marginal increase in the silymarin contents as a result of inoculation. The weight of 1000 seeds from inoculated plants was higher by 14.25% to than the control plants and the amount of the seeds harvested from 10 plants was 25% more than the control plants which will be beneficial to the farmers in taking 25% more production than the control plants. Suring October, 04 the seeds collected from the inoculated and non inoculated plants were sown and the plants raised from the seeds of the inoculated plants were in flowering stage during 3rd week of March 05 while the plants raised from the seeds of the non inoculated plants were under going vegetative growth only. The results will be presented in the Final Report to be submitted in September, 05.

2. *Clitoria ternetea*

One of the interesting observation while doing inoculation of *Clitoria ternetea* was that the leaf area of the leaves of the plants raised from the seeds obtained from the inoculated plants was practically 150 t more than the plants raised from the seeds of the non inoculated plants during monsoon of 2004. The plants after completing the dormancy in Dec-Jan (2004-05) are again sprouting and the leaf area of the plants either inoculated with *Ecorrhiza* or raised from the seeds of the inoculated plants was 100-200 t more than the control plants. More results pertaining to Phytochemical studies and other parameters will be presented in the final report when the data will be available during June 2005 after the flowering is over and seeds are harvested.

3. *Cassia occidentalis*

The seedlings of *Cassia occidentalis* raised from the seeds collected from the plants growing in the campus of Sri Venkateswara College, were inoculated both in the pots and the beds in an effort to undertake the field trials. Like the results obtained with other plants, the growth of the inoculated plants was much faster than the non inoculated plants. The average collar girth of the inoculated plants was found to be 92% more than the non inoculated plants during December 2004. However, when the measurements were undertaken during March-April, 2004, the collar girth was found to be only 13.20% higher in the inoculated plants.

The increase in the height of the plants was noted during the vegetative growth period in June and July 04. there was a increase of 8.53% in the height of the inoculated plants in mid June while 30.62% increase in the height of the inoculated plants was observed during mid July 04. The Collar girth in the inoculated plants was higher by 30% than the non inoculated plants, more seedpods were formed in inoculated plants and the quantity of the seeds collected from 10 pants was 35% more than non inoculated plants. The leaves and the seeds of the non inoculated and inoculated plants were subjected to the total anthraqinane contentes and the inoculated plants were having higher % of the anthraquinone both in the seeds and the leaves in comparison to the non inoculated plants.

Totalanthraquinones: Seed, wild:0.088+0.006, Control:0, 084+0.001, VAM:0.103+0.003: Leaf:Cont.0.067+0.007, VAM: 0.079+0.005,

4. *Acalypha indica*

While doing the inoculation of *Ecalypha indica*, some of the interesting observations are as follow:

i. The growth of the inoculated plants was much faster
ii. The leaf viral attack on the inoculated plants was very less
iii. The total height of the inoculated plants was 15-20% more than the non inoculated plants
iv. The flowering was early
v. The branching in the inoculated plants was more than the non inoculated plants.

The total biomass in the inoculated plants was higher by 20% than non inoculated Plants. The average height of the plants inoculated with *Ecorrhiza* (taken for 8 plants) as observed on 12.6.04 and 17.7.04 was found to be 39.75 and 78.63 cm for control plants and 47.41 and 89.5 cm for the inoculated plants (average taken for 12 plants). The increase in the average height in the inoculated plants was found to be 19.27% and 13.86% more at those dates as compare to the non inoculated plants. However, at the maturity stage after 3 months of plantation, the average height of the inoculated plants was 114 cm and non inoculated plant was 110.0 cm.

The total average biomass of the aerial plants was 123.5g (Inoculated) and non inoculated was 101.5g, there was an increase of 42.68% in biomass as a result of *Ecorrhizal* inoculation. The increase in the root weight was 26.47% more in inoculated plants than the control plants. It is interesting to note that the panchang (whole plant) is used in the Indian System of Medicine.

5. *Phyllanthus amarus*

The results of inoculation in *Phyllanthus amarus* are equally encouraging. The inoculated plants were longer in length, collar girth and total biomass. It is significant because, whole of the plant is used in Indian system of medicine. The increase in the biomass in the inoculated plants was found to be 34.3% more than the non inoculated plants. The collar thickness was found to be 70% more in the inoculated plants. However, per unit mass, the % of the active ingredient Phyllanthin (1.7%) was found to be exactly the same. However, per unit area, because there is 34.5% increase in biomass in inoculated plants, one should expect higher yields.

6. *Catharanthus roseus*

The results of the Mycorrihzal inoculation in *Catharanthus roseus* are also equally encouraging. The average collar girth of the inoculated plants was found to be 30% more than the non inoculated plants. The average biomass in the inoculated plants was found to be 92.03% more than the non inoculated plants. The roots and the leaves are with Natural Product Division (Prof. K.K. Bhutani) NIPER, Mohali for the analysis.

7. *Lepidium sativum* and *Linum uritatissimum*

The weight of the 1000 seeds from inoculated *Lepidium sativum* was found to 4.5 and Control Plants was only 3.6g and there was increase of 25% increase in size of the inoculated seeds while in case of *Linum uritatissimum*, the 1000 inoculated seeds were weighing 19.15% more than the non inoculated plants. The oil contents in case of *Linum uritatissimum* is to be analyzed . An increase of 25% in the weight of the inoculated seeds will result in overall higher yield per acre.

8. *Asparagus racemosus*

After the preliminary results of inoculation on the seedlings of *Asparagus racemosus*, the field trials were undertaken both in the college campus and at Surjeevan Farm, Village Besar Akbar Pur, Distt. Gurgaon, where the soils are sandy. The plants were uprooted after 1.5 years of plantation and comparative morphological studies indicated the following results.

Three and four branches of the inoculated and non inoculated plants were taken for the comparative studies. There was 97.94% increase in the average weight of the tubers from the inoculated plants which are of medicinal importance.

In the bunch of the tubers of the inoculated plants, the thickness of the 168 tubers which were obtained varied from 0.6 cm – 24 mm while in the control plants, the 148 tubers which were obtained, the thickness only was in the range 0.8mm to 10 mm. The length of the tubes in the inoculated bunch was in the range 5cm to 44cm while in the control plants, it was in the range 4cm – 33cm.

The tubers harvested after 1.5 year of plantation were dried after peeling off the thin layer and subjected to HPTLC (M/s Natural Remedies, Pvt Ltd) while in the inoculated plants, Shatavarin IV was found to be <0.1 %, shatarvarin was so low in the noninoculated plants that it could not be detected indicating that the *Ecorrhizal* inoculation helps in the easy synthesis of the higher % of the active ingredient.

9. *Andrographis paniculata*

The inoculation of 3 months old saplings was done with *Ecorrhiza* directly in the fields as Surjeevan Farm during Feb'03 and the plants were allowed to grow and the arial parts of the plants were harvested during Nov'04. The inoculated plants in the 5'x20' beds showed good branching, the flowering was earlier by 10 days than the non inoculated plants. The foliage area spread over in the inoculated plants was 20-25% more than the non inoculated plants. The leaves and twigs were analyzed for *Andrographolide* and there was no change in the % of the active ingredient per unit mass as a result of inoculation. However, the biomass which is medicinal value was found to be 26% more in the inoculated plants, the inoculation with *Ecorrhiza* can give higher field and hence may be incorporated in the cultural practices.

10. *Bacopa monnieri* and *Centella asiatica*

Bacopa monnieri, Brahmi is a very significant plant of Ayurved and has also been studies by us. The inoculation studies were carried out at the Surjeevan Farma, Dist. Gurgaon. These were carried out in the fields where the soils are sandy soils. The plants were inoculated with the mycorrhiza inoculum by putting 1.5 g of the inoculum 2" deep round the root zone during the ist week of July and allowed to grow for three months. The plants were uprooted both from the control and inoculated beds. During the growing period, there was enough growth in the inoculated beds to the extent that the soil was not visible and in the control beds, the soil could be seen and it was scanty growth.The % of Bacoside-A3 was found to be 0.65 in the inoculated plants and was only 0.24 in the control plants. Similar results were obtained in *Centella asiatica*, Mandookparni where the % of Brahmic acid in the inoculated plant was found to be 1.49 and only 0.59 in the control plants.

11. *Adathoda vasica*

The cuttings of *Adathoda vasica* were obtained from the plants growing wild in the Aravali Ranges, Village Bhondsi, Distt. Gurgaon, were rooted using 2000ppm sols. of IBA using quick deep method and the rooted cuttings were used for the inoculation studies. These were planted in the control and experimental beds adjacent to one another and allowed to grow for one year. The flowering in the inoculated plants was observed 20 days before the control plants, the plants were harvested , the leaves removed, shade dried and subjected to quantification of Vasicine in the control and inoculated plant leaves. The % of Vasicine was found to be 0.384 in the inoculated plants and was as low as 0.166 in the control plants.

11. *Ocimum sanctum* and *Anrograhis paniculata*

Ocimum sanctum and *Andrograhis paniculata* results at the Reseach Farm , Natural Remedies, Bangalore.Very recently, an experiment was conducted at the Research Farms of Natural Remedies, Bangalore to assess the difference in the yields and chemical contents of *Andrograhis paniculata* and *Ocimum sanctum* on our request to Dr.Amit Aggarwal, Director Research, Natural Remedies, Bangalore. 100g of the *Mycorrhiza* inoculum was mixed with thoroughly with 2 kg of the organic compost obtained from the compost garden. The experiment was conducted at the Research Farm situated near Denkanikota in Tamil Nadu. This was irrigated land located at 900mts above sea level and is free from the pesticides and chemical fertilizers, with an average rainfall of 900 mm per year. The soil is of red variety and overall the soil is good for growing *Andrograhis paniculata*.

Two experimental beds of 10X10 ft were made , one of which was selected for the experimental and the other one served as control. Seedlings of *Andrograhis paniculata* which were 6 weeks old were planted at a distance of 20to 30 cms apart. The bed was maintained for a period of 3 weeks. A small groove of 4 inch depth was made 2 inches away from each row and of the plant in the experimental bed. The inoculum was applied uniformly in the groove and the groove was covered with the soil mixture.These beds were irrigated at regular intervals and care was taken such that the water did not flow from one bed to another. Two to three rainfalls occurred during the experimental period. The beds were also maintained free from the weeds. No pesticides and chemical fertilizers were applied during the experiments. Kalmegh plants grew to an average height of 30-40 cms in a period of 3.5 months after which they were harvested and dried. The sample was submitted for HPLC analysis. The results are tabulated below.

Results: Kalmegh.

Period of Experiment : 111 days.

Date of transplantation : 02.09.08

Date of *Mycorrhizal* application:25.9.08

Date of Harvest; 22.12.2008.

	Yield (Wet)	Andrographolide By HPLC	Water soluble extractive	Alcohol soluble extractive
Control	1.940 kg	1.47%	25.25%	13.23%
Treated	2.450 kg	1.51%	23.43	13.14%

Based on the above experiment, it is concluded that there was a considerable increase (12.5%) in the biomass in the trated plants over the control group of plants. However, the increase in the Andrographolide content was found to be insignificant. The increase in the biomass however will be beneficial to the farmers in terms of the more production and higher profits.

Results on *Ocimum sanctum*:

Tulsi

Duration of experiments; 107 days

Date of transplanting; 04/09/08

Date of *Mycorrhizal* inoculation: 25/09/2008

Date of harvest: 20/12/2008

	Yield (Wet)	Ursolic acid	Water soluble extractive	Alcohol soluble extractive
Control	5.710	0.645%	27.42%.	12.77%
Treated	6.550kg	0.63	24.78	14.42%

Though there was insignificant increase in the Ursolic acid content as a result of inoculation, the biomass which is of commercial value , increased to 20.81%.

Research Results from Centre for Mycorrhizal Research (CMR), The Energy and Resource Institute (TERI), New Delhi.

The scientists of the Centre for Mycorrhizal Research , TERI, New Delhi have also carried out extensive studies of the *Mycorrhizal* inoculation on *Jatropha curcas* during the last five years and have developed an unconventional method of growing the plant faster and better. It was during these studies it was observed that the root infection of the plants are practically 100 % because of which *Jatropha curcas* is used as host plant for the trap culture studies.The CMR has raised *Jatropha curcas* with *mycorrhizal* application across the country inorder to observe how the plant behaves to the inoculation in the different climatic zones of the country. During these studies, it was concluded that there is high degree of dependence of the growth on the inoculation. There are different methods which are used for the propogation of *Jatropha* and these are very slow. The standard seedling method takes practically two years for the plant to start yielding the fruit. Under the climatic condition of Distt. Gurgaon, where the soils are sandy, porous and the root spread is faster, the flowering starts in the same year if the plantation is done in Feb. and one can get about 150-200 g of the seeds and the subsequent years, it increases because the growth of the plants is comparatively higher than other areas. The plants raised from the clonal culture takes a year for the first yield. If the saplings are inoculated with *mycorrhiza* inoculum at the time of the plantation in the fields results in the hastening the process of flowering and fruiting. The first yield was obtained after seven months of the plantation. This technique also results in upto 30% higher yields and plant biomass which on decompostion during the winter months adds to the soil fertility. The trials were carried out in the seven different agro climatic zones in the country. The CMR's *mycorrhiza* application speeds up the process. The first yield arrives after seven months of cultivation and the inoculation with *mycorrhiza* also increases the yield upto 30%. Wheas the conventional methods *Jatropha* plantations yielded 200 kg per hectare, compact plantation in the wastelands produced 325 kg seeds per hectare in the first year of production using the *mycorrhizal* inoculation technology. Using the technology, *Jatropha* plantations were also established in the industrially created wastelands, for example, flyash overburdens, chlor alkali-laiden sites, sites with distillery effluents and so many other types of the useless lands which can not be used for the plantation programs.

More than 6, 00, 000 plants of *Jatropha curcas* have been established using the *Mycorrhiza* inoculum, which are spread over in the seven different agroclimatic zones across the country comprising the Chhattisgarh plain zones (at Korba, MP), arid western plain zone (Jodhpur, Rajsthan), Malwa Plateau zone (Barwaha, MP), south Gujrat zone (Mithapur, Gujrat), coastal saline zone (Titagarh, West Bengal), Vijayawara and Machalipatanam (south zone of Andhra Pardesh), Kanjivaram (south zone of Tamil Nadu, and many more are planned in the near future. The net results of these plantation programe undertaken in the different climatic zones has contributed to the knowledge on early flowering and high yielding *Jatropha* plants from superior germplasm and on application of the *mycorrhizal* biofertilizer. These results should be taken up by the various NGOs, research workers, people involved in the extension programs and the forest departments who are raising the saplings and undertaking the plantation programs. It is also advised that the forest department human resouse who are involved in the raising of the saplings should in fact raise the *mycorrhized* saplings and the mycorrhiza inoculum be mixed with the potting mixture @ 2kg per 300-400 kg of the soil/manure mixture and the seeds can be directly sown in the polybags. A additional dose of the inoculum @ 2g per micro pit may be put at the time of the plantation of the one year old saplings. The mycorrhized saplings will be already having the mycorrhiza spore and further infection in the newly developing hairy roots will boost the growth. The additional cost involved per plant is only 30-40 paise interns of the cost of the inoculum but you get 30% higher yields not only in terms of the biomass but also in terms of the yield of the seeds and finally the oil.

HOW TO USE THE MYCORRHIZA INOCULUM

Mycorrhiza inoculum is available in 1 kg packet which is sufficient for 700 plants, At the time of the plantation of one year old saplings , 1.5 g of the *Mycorrhiza* inoculum which is available either in the fly ash or in the lignite as a carrier may be added in the micro pit which is of the size of the earth ball of the plant, is mixed well and the plant sapling is inserted in the micropit along with the earth ball , pressed and irrigated lightly. Flood irrigation may be avoided at the time of plantation. After 5 days of the inoculation, the results will be visible.

RAISING THE *MYCORRHIZED* SAPLINGS

Mycorrhized saplings can also be raised either in the polythene bags or in the nursery beds. *Mycorrhiza* inoculum be directly mixed in the potting mixture at the rate of 1kg of the inoculum per 500 kg of the soil mixture and seeds may be sown. If the saplings have to be raised in the beds, *mycorrhiza* inoculum be mixed in the beds @100g of the inoculum per 500 sq feet area. It may be mixed with the cow dung /vermicompost (@ 20 kg) and mixed in the bed uniformly so that the spores are distributed uniformly and are made avable to each and every plant for better growth.

The sources to get the *Mycorrhiza* inoculum.

1. Majestic Agronomics. Pvt Ltd.
   Regd. Office :C-1027, Palam Extension, Sector-7 Dwarka, New Delhi-110045,
   Works: Mahat Pur Industrial Area, Mehat Pur, Una, Distt.Una Himachal Pardesh. anupsharma1154@hotmail.com, Cell:09811105153
2. KCP Sugar and Indusries Corporation Ltd,
   Vuyyuru-521165 Krishna Distt, AP
   Email: kcpsugar@vsnl.com, vjwkcpvyrnd@sancharnet.in
3. Cadila Agro Division,
   Corporate Campus, Sarkhej Dhola Road, Bhat, Ahmedabad-382210

RESEARCH CARRIED OUT AT UTTHAN, ALLAHABAD

Majestic Agronomics Pvt Ltd in association with R &D Unit of People for Plants and Environment (PFPE), an NGO of Academics, Environmentalists and Horticulturists, have initiated research studies on the *mycorrhizal* inoculation of *Jatropha curcas* in the trial fields of the well known NGO- UTTHAN in Allahabad. The inoculation of the plants were done very recently in May'06 at three different locations . The three sites selected were the degraded soils. One was where the Bric kilns were operational at one time and after removing the 4-6 feet rich soils, the site was left with nothing but the deficient soils and the other site was which is the calcarius soils not fit for any type of the farming. In comparison to the noninoculated plants, the inoculated plants have shown better growth, the leaves of the inoculated plants were having more shine because of the greater photosynthesis, It is interesting to note that the inoculation which was done in May'06 of the plants which were one year old saplings in the fields resulted in the growth at par with the plants which were two years old without inoculation. The dark green colour in the leaves of the inoculated plants make them easily spotted for a different group of the plants. It is also interesting to note that the inoculated plants needed lesser water but still showed better growth while the two years noninoculated fields , the plants were less greener and pale yellow The plants are under observation for the various morphological parameters. But we are expecting similar results as obtained by the scientists of TERI, New Delhi.

FURTHER RESEARCH ON THE INOCULATION OF *JATROPHA CURCAS*

The author is busy in the research on the response of the inoculation of *Jatropha curcas* with various microbial inoculants where our main target is to see how this inoculation affects the growth of the plant and the effect on the oil yield and the various constituents are affected or not. Shortly we will come out with the results.

RESEARCH ON HIMALAYAN MEDICINAL PLANTS

We have initiated our investigations on the RET category medicinal plants of the Himalayas in association with a reputed Institute of Uttrakhand, SHER- Society for Himalayan Environment and Research to investigate the effect of the *mycorrhizal* inoculation of these medicinal plants on the bioactive phytoconstituents.

REFERENCES

Bhoon, Yudhvir K. and Prasad, N.V.S.R.K. Comparative inoculation studies on *Oxalis cernea* with *Piriformaspora indica* and Effective microorganism. *Indian J. Microbiology* Communicated.

Bhoon, Yudhvir K. and Prasad, N.V.S.R.K. Effect of Arbuscular Mycorrhizae (AM) and Effective Microorganism (EM) inoculation in Medicinal Plants 1. *Solanum nigrum, Indian J. Microbiology* Communicated.

Bhoon, Yudhvir K. and Prasad, N.V.S.R.K. *Vesicular Arbuscular Mycorrhiza* inoculation of an important medicinal plant. *Asperagus racemosus* Wild : Pot and Field Studies. *Indian J. Microbiology*, Communicated.

Kumari, R., Pham, G.H., Singh, A., Bhoon, Y.K., Srivastava, A.K. and Varma, A. 2004. Biotechnological Processes for Transfer Technology of Medicinal Plants: Mediated by Symbiotic Microorganisms (From Lab to Fields Novel Concepts). World Herbo Expo, Bhopal.

Kumari, Rina , Krishan, Hari, Bhoon, Y.K. and Varma, Ajit, 2003. Colonisation of cruciferous plants by *Piriformaspora indica. Curr. Science*, 85: 1672-1674.

The work reported in the chapter is based on the Final Report submitted to NMPB, incorporating the results of our investigations in the Research Project No. 87 "Biodiversity of AMF in the Medicinal Plants of Aravali Ranges Distt Gurgaon and Reserved Saraswati Forest, Distt Kaithal, Haryana; inoculation of the Medicinal Plants with Mycorrhiza and Comparative Phytochemical Investigations" sanctioned by NMPB, MHFW, GOI to the author from 2002-05.

Chapter **13**

Biogas Manure (BgM) : A Rich Source of Plant Nutrients and Substitute to Inorganic Fertilizers

S.K. KHANDELWAL, G. RAJAMANI AND B.K. MISHRA

Department of Molecular Biology and Biotechnology, Rajasthan College of Agriculture, (Maharana Pratap University of Agriculture and Technology), Udaipur (Rajasthan) 313 001

INTRODUCTION

In modern agriculture, the requirement of major plant nutrients is primarily met through application of inorganic fertilizers. World agriculture has now reached a stage that chemical fertilizers can not be replaced by achieving higher and continued productivity of the crops to meet the growing food requirement. It has been reported that average fertilizer consumption in India increased from 0.5 kg/ha during 1951-52 to 117.07 kg/ha during 2007-08 (**Economic Survey, 2008-09**). Intensive agriculture with the use of agrochemicals in large quantities has increased the land productivity of various farm commodities manifold. Nevertheless, adverse effects of these chemicals are clearly visible on soil microflora, quality of water, food, fodder etc. Pesticides and nitrates from fertilizers have been detected in ground water of many agricultural regions. High concentration of nitrates in drinking water affects human health.

With an exponential increase in population each year, countries are compelled to increase agricultural food production in a sustainable manner. Addition of organic manures like farm yard manure (FYM), compost, crop residues, biogas plant spent slurry (biogas manure or BgM), vermicompost, etc. help in improving soil fertility on a long term basis. Such soil amendments also change rhizosphere environment by affecting soil porosity, aeration, temperature, water holding capacity and soil microflora. Under **National Biogas Manure and Management Programme** (NBMMP), which is pursued by Ministry of New and Renewable Energy, Government of India, New Delhi, approximately 4.17 million biogas plants **(Akshay Urja, December 2009)** have so far been constructed primarily in the country side. These plants annually generate approximately 30.18 million metric tonnes of biogas plant spent slurry as by-product that is normally used as nutrients to supplement crop production.

The NBMMP continue to operate in the 11th Five Year Plan in all States and Union Territories. The programme forms part of the Government of India 20-Point Programme, the objectives of which are as follows-

- To provide clean bio- gaseous fuel mainly for cooking purposes and also for other applications for reducing use of LPG and other conventional fuels;
- To meet 'lifeline energy' needs for cooking as envisaged in 'Integrated Energy Policy';
- To provide bio-fertilizer/ organic manure to reduce use of chemical fertilizers;
- To mitigate drudgery of rural women, reduce pressure on forests and accentuate social benefit;
- To improve sanitation in villages by linking sanitary toilets with biogas plants;
- To mitigate Climate Change by preventing black carbon and methane emissions.

WHAT IS BIOGAS MANURE (BGM)?

BgM is a by- product obtained from the biogas plant as a result of fermentation of cattle dung or other biomass, yielding generation of methane rich gas. BgM contains a full range of plant nutrients. It's use as organically enriched fertilizer gives farmers a double advantage of having installed their biogas plants. The fibrous material, inorganic solids that don't get digested or converted into methane either settle down in the plant or come out with slurry liquid through outlet. This contains many rich and nutritive elements including nitrogen, phosphorus, potassium and trace elements i.e. Zn (144 ppm), Mn (188 ppm), Fe (3550 ppm) and Cu (28 ppm). BgM improves physical, chemical and biological properties of the soil that in turn enhances crop yields.

The nitrogen present in cattle dung is conserved totally when processed through a biogas plant whereas in open pit composting most of the nitrogen (over 50%) is lost due to leaching or evaporation. About 20% of the total nitrogen present in BgM is in ammonical form which is readily available for crops. However, this form of nitrogen is rapidly lost if the digested slurry is sun-dried.

BgM nourishes the soil with supply of essential nutrients to accelerate the growth of plants, especially for root growth due to humus and aeration which enhances crop yield in a sustainable manner. Moreover, during anaerobic digestion weed seeds gets destroyed. As a result, BgM is free from weed seeds and there is no foul smell from the fields. The application of BgM enhances the fertility of the soil to optimize quality production. Due to its dark colour, it absorbs sunlight thereby resulting in warming up of the soil. Following are the advantages of BgM-

- It has no bad smell.
- Insects do not grow in it.
- Micronutrients in soil increases, thus improving soil structure.

- Water holding capacity of the soil is enhanced that ensures plant growth.
- Retains and enhances nutrient uptake for better nourishment of plants.
- Enhances the aeration of soil for root penetration resulting in better growth.
- The viability of weed seeds is lost.
- Less methane exposure.
- Qualitative improvement in taste, smell, size and colour of the food product etc.
- Readily available, can be used whenever needed.

BGM PLANT MODELS

Approved models of family type BgM plants—Appropriate models of family type biogas plants should be selected on the basis of preference of the beneficiaries and considering technical requirements, such as location, distance between kitchen and cattle shed, availability of water and feed-stock like dung, loose and leafy biomass, sanitary and other biomass wastes. Approved models for such plants are available for 1 to 10 cubic metre capacities for fixed dome and floating dome type plants. The commonly used capacities of these models are 1-4 cubic metres. Pre-fabricated models of biogas plants are also available based on High Density Polyethelene (HDPE), Fiber-glass Reinforced Plastic (FRP) and Reinforced Cement Concrete (RCC) material in addition to ferro-cement and brick-masonry biogas plants (Table 1).

Table 1 : Approved models of family type biogas manure plants

S. No.	Biogas fertilizer plant models*	Specifications/ Ministry's letter No. and date
1.	**Pre-fabricated biogas plant models:**	
	i) Pre-fabricated RCC fixed dome model.	Ministry's letter No.13-10/96-BG dated 10-1-2002.
	ii) Pre-fabricated RCC digester KVIC model family type Biogas Plants.	Ministry's letter No.13-1/2007-BE dated 29.2.2008.
	iii) Pre-fabricated HDPE material based complete Deenbhandhu Model Biogas Plants.	Ministry's letter No.13-1/2007-BE dated 3.09.2007.
	iv) Pre-fabricated BIOTECH make FRP Biogas plants.	Ministry's letter No.13-1/2007-BE dated 29.2.2008.
	v) Pre-fabricated HDPE material based KVIC type floating dome Biogas Plants.	
2.	**Floating Dome Type Biogas Plants:**	
	i) KVIC floating metal dome type Biogas Plants.	Code of Practices (Second Revision), IS 9478:1989 of the BIS, New Delhi.
	ii) KVIC type plant with Ferro Cement digester and FRP gas holder.	Code of practice IS-12986:1990 of BIS,New Delhi.

contd. ...

... contd.

S. No.	Biogas fertilizer plant models*	Specifications/ Ministry's letter No. and date
	iii) Pragati Model Biogas Plants.	Code of Practices (Second Revision), IS 9478:1989 of the BIS, New Delhi.
3.	**Bag Type Biogas Plants:**	
	Flexi model	Ministry's letter No.7-39/89-BG dated 14.7.95
4.	**Fixed Dome Biogas Plants:**	
	i) Deenbandhu Model with Brick masonry.	Code of Practices (Second Revision), IS 9478:1989 of the BIS, New Delhi.
	ii) Deenbandhu ferrocement model with in-situ technique.	Ministry's letter No.13-11/99-BG dated 5-3-1999
	iii) Prefabricated HDPE material based prefabricated dome for Deenbandhu Model family size Biogas Plants.	Ministry's letter No No.13-1/2007-BE dated 3.9.2007

*** Newer models would be added depending on technology development.**

Comparison of BgM with other organic manures

Organic manures are valuable by-products of farming and allied industries derived from both plant and animal sources. They can be broadly grouped into two main categories (i) Bulky organic manures and (ii) Green manure crops.

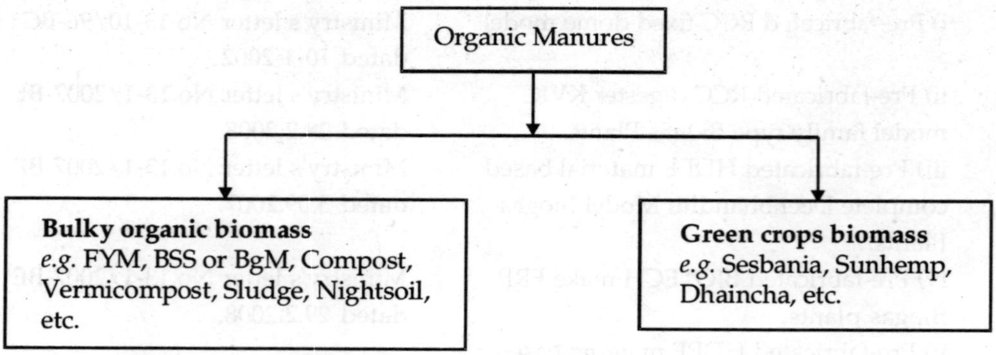

The influence of organic manures on soil quality is as follows-

- Organic manures contain both major and minor nutrients in small quantities. These manures are applied in large quantities per unit area and have a direct effect on plant growth.
- Organic manures increase organic matter content of the soil and therefore, improve its physical properties and water holding capacity.
- Organic manures also stimulate microbial population which in turn helps make nutrients available to the plants.

Farm yard manure (FYM)—This term refers to the decomposed mixture of dung and urine of farm animals along with litter (bedding material) and left over material from roughage, fodder, etc. fed to the cattle. FYM consists mainly of dung and part of urine soaked in the refuse. The average composition of fresh excreta and urine of various animals is given in Table 2. In general, urine of all animals contains higher nitrogen and potash as compared to dung.

Table 2 : Average nitrogen, phosphorus and potash contents of fresh excreta of common farm animals

Animal(s)	Excreta	Nitrogen (N_2O)	Phosphorus (P_2O_5)	Potassium (K_2O)
Cow and Bullock	Dung	0.40	0.20	0.10
	Urine	0.75	Trace	1.35
Sheep and Goat	Dung	0.75	0.50	0.45
	Urine	1.35	0.05	2.10
Horses	Dung	0.55	0.30	0.40
	Urine	1.35	Trace	1.25
Pigs	Dung	0.55	0.50	0.40
	Urine	0.40	0.10	0.45

Green Manure—Green manuring can be defined as a practice of ploughing or turning into the soil undecomposed green plants for improving physical structure and fertility of the soil. It is an age-old practice followed by the farmers to improve soil fertility.

The main advantages of the green manuring are-

- It stimulates the activities of beneficial soil microorganisms by addition of organic matter.
- It improves the soil structure and reduces runoff and soil erosion.
- Green manuring with leguminous crops (Sunhemp, Dhaincha and Sesbania etc.) also adds nitrogen to the soil through biological nitrogen fixation.
- They also help in conserving soil nutrients by avoiding leaching loss.

The success of green manuring depends upon proper availability of moisture at decomposition sites. Besides, farmers practising green manuring should be aware of possible pests and diseases of such green manure crops and their impact on successive plantations.

Compost—Composting is largely a biological process in which microorganisms of both types, namely, aerobic (which require air or oxygen gas for development) and anaerobic (which function in absence of air or free oxygen) decompose organic matter and lower the carbon-nitrogen ratio of refuse. The final product of composting is well rotted manure, known as compost. Because of the high organic matter content and due to a relatively higher content of major nutrients compared to farmyard manure, compost is valuable for improving and maintaining soil fertility in crop production.

Compost is prepared from vegetable and animal refuse collected on the farm or in towns or villages. As such composts differ in nature and composition. Composts commonly prepared in India are of the following nature:

- **Rural compost (Compost from farm litter):** This utilizes weeds, crop stubble, bhusa or straw, crop residues such as cotton stalks, groundnut husks, leaves, sugarcane trash, urine soaked earth and litter from cattle shed, waste fodder, hedge clippings etc.
- **Urban compost (Compost from town-refuse):** The main components of town compost are night soil and street and dustbin (house) refuse, commonly known in towns as katchara and wood ashes.

It is estimated that the major sources of compost available in India, like sewage, rural wastes and urban wastes can help in mobilization of 4.064, 1.166 and 2.602 million tonnes of N, P_2O_5 and K_2O annually and taking nearly 50 per cent as rate of utilization, the maximum saving on this account may come upto 2.032, 0.583 and 1.301 million tonnes of N, P_2O_5 and K_2O, respectively. There are at present innumerable difficulties and social prejudices in collecting and processing all these wastes.

The difference between composting and BgM are given in Table 3.

Table 3 : Difference between composting and BgM

S.No.	Composting	Biogas manure
1.	Unselective degradation by both aerobic and anaerobic microbes causing more weight loss without proportional enrichment of fertilizer value. Both carbon and nitrogen are consumed during degradation.	Selective anaerobic digestion. Weight loss is less. Mainly carbon is consumed producing methane.
2.	Proper degradation time is 90-120 days.	Proper degradation time is 30-40 days.
3.	About 30-50 % nitrogen is lost due to evaporation of ammonia.	Loss by evaporation is negligible.
4.	Compost has bad odour.	Odour is minimized.
5.	Some quantity is blown off by wind.	No such loss.
6.	Nitrogen is less.	N, P and K is comparatively more.
7.	Compost has limited macro- and micronutrients.	Biogas manure has more macro- and micronutrients.

Vermicompost—Earthworms are regarded as farmers' natural friends in the soil. They burrow the soil continuously and turn it alongwith the organic material. They feed on this organic material, degrade it and the excreta are referred to as *vermicompost*. The degradation of organic matter in the body of the earthworm is carried out continuously by the microflora present in their intestine. The microflora comprises of several bacteria and actinomycetes that produce enzymes required for the said decomposition. They also produce several vitamins and growth promoting substances that accelerate plant growth.

The main advantages of vermicompost are-

- Fresh vermicompost contains major nutrients (nitrogen, phosphorus and potash) and micronutrients (0.16% zinc, 0.03% copper and 1.38% iron). Its pH normally varies between 6.8-7.2. Every quintal of fresh vermicompost can supply approximately 800g, 1100g and 500g of N, P and K besides several trace elements which are easily made available to the plants.

- Vermicompost improves aeration and water holding capacity of the soil. It also reduces soil erosion.

Biogas plant spent slurry (BSS or BgM)—The slurry coming out of biogas plant is referred to as biogas plant spent slurry. The digested slurry discharged from the common biogas plants normally contains 92-94% moisture. It is rich in N, P, K and several micronutrients. BgM improves physical, chemical and biological properties of the soil that in turn enhances crop yields. BgM can be used as a rooting medium for horticultural crops and it also supports the growth of aquatic fern plants and algae Azolla.

Anaerobic fermentation of dung or other biomass in biogas plant reduces the C/N ratio by removing some of the carbon, which has the advantage of increasing the fertilizing effect. Another favorable effect is that nitrogen and other plant nutrients become mineralized and hence are more readily available to plants. More over, well digested slurry is practically odorless, easier to spread and does not attract weeds and insect flies. Table 4 indicates how BgM is more effective than other organic manures in relation to NPK content.

Table 4: Comparison of biogas manure with other organic manures (Average Percentage)

Manure	Nitrogen (N_2O)	Phosphorus (P_2O_5)	Potassium (K_2O)
Fresh cattle dung	0.3-0.4	0.1-0.2	0.1-0.3
Farmyard manure	0.4-1.5	0.3-0.9	0.3-0.9
Compost	0.5-1.5	0.3-0.9	0.8-1.2
Biogas manure (BgM)	**1.5-2.5**	**1.0-1.5**	**0.8-1.2**
Poultry manure	1.0-1.8	1.4-1.8	0.8-0.9
Cattle urine	0.9-1.2	Trace	0.5-1.0
Paddy straw	0.3-0.4	0.8-1.0	0.7-0.9
Wheat straw	0.5-0.6	0.1-0.2	1.1-1.3

Chemical fertilizers and BgM—The inorganic chemical fertilizers are harmful in the long run as they do not provide balanced diet to plants, severely affecting the physical, chemical and microbial properties of the soil. The extensive use of inorganic fertilizers result in following impacts-

- Destroys soil microflora, especially the nitrogen fixing bacteria.
- Causes pollution of fresh water reserves.
- Reduces soil porosity, aggregation and ultimately to infertility.
- Erodes topsoil due to wind because of missing organic matter in the soil.
- Is not cost effective on long term basis.

Table 5 indicates the differences in the uses of BgM and the chemical fertilizer urea in general.

Table 5 : Differences in the uses of BgM and chemical fertilizer (Urea)

S.No.	Biogas manure	Urea
1.	Balanced nourishment	Imbalanced nourishment
2.	Pollution free	Causes pollution of water, air etc.
3.	Eco-friendly	Damage to the ecology
4.	Defense against pests	Vulnerable to pests
5.	Can be made at home	To be bought from outside only
6.	Cost effective	Fluctuating cost
7.	Needs less water	Needs more water
8.	Maintains soil fertility	Spoils the soil strength
9.	No side effect	Many adverse effects
10.	Quality food	Less healthy food
11.	Better health	Poor health
12.	Wholesome	One sided
13.	Less yield but sustainable returns	More yield and quick returns
14.	Increasing returns	Diminishing returns

The study of BgM in the crop maize has resulted in improving the plant parameters (Table 6).

Table 6 : Comparison of impact of urea and biogas manure in crop parameter of maize

Treatments	Plant height (m)	Plant dry weight (g/plant)	Cob weight (g/cob)
Urea	2.10	216	188
Biogas manure	2.14	249	263

BgM is also found to be a very good nutrient input to the horticultural crop tomato (*Lycopersicon esculentum*) (Table 7).

Table 7 : Comparison of fruit yields in tomato (*Lycopersicon esculentum*) when treated with BgM

S. No.	Treatment	Fruit yield (kg/plot)
1.	Biogas manure	9.9
2.	Biogas manure with urea	8.6
3.	Control	7.5

The demand of organic manure can be quickly met by recycling of available organic wastes from households, dairy, agro-industries, food processing industries and crop residues. The anaerobic digestion through recycling not only provides sanitation but gives biogas and enriched manure. Therefore installation of biogas plants to utilize animal dung (from any source) has dual advantages as described above.

Utility of BgM—Biogas manure has wide applications besides enriching the soil as crop manure. It can be used for fish culture as well as mushroom cultivation to increase production of fish and mushrooms. BgM can also be used to enrich inorganic fertilizers, vermicompost, NADEP compost, farmyard manure etc. Cattle urine can also be used with BgM as diluents.

Manuring crops—BgM can be used effectively in any type of soil and for any crop. The recommended application of manure is 10 tonnes per hectare in irrigated areas and 5 tonnes per hectare in dry land areas. However, the response of crops would vary. The increase in crop yield is about 10 to 40 %. Most responsive amongst crops are vegetables and fruit trees, particularly root crops such as carrot, radish, sweet potato, tea, coffee, sugarcane, rice and jute. BgM can be used as fertilizer for crops in any one of the following forms:

1. Liquid BgM—Maximum benefit is obtained when the slurry is used in liquid form as it comes out of the biogas plant. It can be easily distributed in the fields if the slurry outlet of the plant is linked with the main irrigation channel. This would be possible only if the biogas plant is installed on the farm. Otherwise, BgM can be transported in discarded oil drums or buckets, by wheel barrow or bullock cart, which is more cumbersome. BgM has a solid content of about 6 per cent, a pH value of about 8 to 9 and nitrogen 1.8 % along with other nutrients. This is the best form of BgM for application because-

- Has better nitrogen composition compared to dry and semi-dried slurry.
- As a soil conditioner. Aggregated soil communicates absorption of the slurry. Moreover, the bacterial and fungi growth is enhanced on application of BgM, which is crucial for plant and crop yield.
- Good for acidic soils.
- Reduces harmful elements like aluminium thereby minimizing toxicity.
- Supplies nutrients to beneficial soil microbes.
- Changes membrane permeability of root hairs and enhances nutrient uptake.
- The water holding capacity of the soil increases.

2. Semi-dried BgM—This is the next best form for use. This form of BgM may be spread in shallow pits and allowed to dry partially. It is then scraped and stored in piles which should be covered with plastic sheets or mats until applied to the fields. It has a solid content varying from 15 to 20% and pH 7 to 9.

3. Dry BgM—It is the slurry coming out of the plant and stored in open for some time period before application. The content of solids varies from 20 to 30 per cent and the pH value from 7 to 8. Dry slurry has micronutrients but almost 20% less nitrogen as it gets lost due to sun drying. Complete drying under the sun should be avoided.

The wet BgM if not required for direct application, can be stored in manure pits, normally made of 1x1x1 to 2x2x1 m size along with dry fodder, crop residues etc. kept for 3-4 months. It is not kept more than 1 m deep in order to avoid handling problem. The dry biogas manure is transported and spread in the field uniformly before ploughing. The BgM mixes well in the soil during ploughing and available to roots uniformly throughout the depth of the field.

Pre-treatment of BgM before application—The slurry from sewage sludge and night soil based biogas plant contains pathogens as such which are not fully destroyed during digestion, if retention time is kept less and also if digester temperature is lower than 35° C (Table 8). It has to be pre-treated before application. Therefore, killing *Ascaris Ova* and

Colitire before use of BgM is necessary, for which the slurry should be treated with ammonia or exposed to open sun dry at 30-35°C, for 2-3 days, otherwise the pathogens may be present in fruit and vegetable crops and have harmful effect.

The BgM of night soil based biogas plant is stored in a heap or in drying pits; some amount of urea is added to it and kept covered to kill the harmful pathogen. The ammonia produced from the urea destroys the pathogens.

Table 8: Pathogens in night soil BgM

Types of pathogens	Minimum retention time in days	Fatality per cent (35° C – 37° C)
Colitire	21	10^4 number
Schisto-some Ova	07	100
Hookworm Ova	10	100
Ascaris Ova	36	98.8

The sewage sludge contains plant nutrients and humus forming effect and as such it is treated as valuable organic manure. However, the sludge contaminant accumulates in it during digestion process such as heavy metals, additives (DDT, detergents, etc.), organic pollutants, etc. in such quantities which prevents its direct use in agriculture. Otherwise, these may concentrate in the field and effect soil fertility adversely. The Table 9 shows the level of contaminants typical urban sewage sludge.

Table 9 : Contaminants in sewage sludge

Metals/Compounds	Quantity (mg/kg)
Heavy metals	
Lead	300-500
Cadmium	5-15
Chrome	100-300
Copper	500-800
Nickel	50-150
Quick silver	~5
Zinc	2000-4000
Organo-Chlorinated Compounds	
Hexa chlorobenzene	0.01-0.2
Dioxin and Furan	40-750
Tetra chloroethylene	1-50
Other Organic Compounds	
Polycyclic aromatic hydrocarbons	1-6
Nonyl Phenol (1-10)	1-1000

Therefore, sludge slurry is applied after tests and later ensured that its utilization keep field free of contaminant deposits or after toxic materials are removed before use. The use of pesticide on sludge applied field build up contaminants in the soil.

Table 10 gives the limits of heavy metals in air dried soil which can be tolerated in soil provided they don't concentrate and pH of soil is not less than 5.

Table 10 : Desirable limits of heavy metals in soil

Heavy metal	Light soil(mg/kg)	Other soil(mg/kg)
Lead	100	100
Cadmium	1	2
Chrome	100	100
Copper	60	100
Nickel	30	50
Quick silver	1	1.5
Zinc	150	300

The organo-halogen compounds should not exceed 500 mg/kg of dry mass and it is not recommended to apply more than 5 tonnes of dry sludge per ha in a period of 3 years.

Effect of use of BgM—The effect of BgM has been studied at many places by several scientists and it is reported that use of BgM increases the crop yield in comparison to FYM. It is also reported that quality of fruits or vegetables was superior in comparison to FYM.

Table 11 shows the results of application of BgM and FYM on yield of different crops.

Table 11 : Comparison of yield from application of FYM and Biogas manure

S. No.	Crop/ Vegetables	Yield Quintal/ha	
		FYM	Biogas manure
1.	Mustard	54.61	60.41
2.	Gram	8.75	11.03
3.	Berseem	196.66	233.32
4.	Oat	221.66	234.00
5.	MP Cheri	30.55	329.70
6.	Cabbage	77.96	96.00
7.	Tomato*	187.00	198.00
8.	Potato*	13.06	20.00

*** Good in size and quality.**

The weed population in slurry plots was much less than FYM plots; therefore use of biogas plant slurry or BgM reduces the cost of labour involved in cleaning the weeds from the field.

The regular analysis of soil was also done to study the residual effect of biogas manure for six crops. The amount of nutrients available to plants is more in case of BgM treated than FYM. However during subsequent years, the available nutrients were same in both the cases. This shows that BgM should be applied during every crop and at regular intervals.

Other uses of BgM—The BgM has many other applications, as in pisciculture, wormiculture, algae and mushroom production, etc.

1. Pisciculture—BgM has been used in fish ponds resulting in healthy growth of fish, control of common diseases and improved fish quality. It is concluded that addition of BgM increases the level of dissolved oxygen in pond water thus helping in more activities and increase in weight of the fishes. Moreover, BgM does not contain harmful bacteria, pathogens and media for growth of disease causing bacteria and thus the incidence of diseases in fishes reduced from 60-70 % to 5 %. Also, BgM helps in quick growth of algae, a food of liking of fishes that makes available additional nitrogen. This enhances free amino acids content of proteins, particularly lysine. In experiments conducted in India and China, it has been reported that the use of biogas manure increased the quantity of produce and reduced the harvest period. Application of BgM in a 0.1 ha of pond area resulted in 762 kg of fish production in a time period of 11 months. This is a commendable performance without making use of artificial feed. Similarly, two types of fish (Catla and Silver) were raised in six months showing record weights of 530g and 900g, respectively. The quantity of application of slurry depends on the oxygen budget of ponds, fish density and the population of aquatic plants. Utilization of digested slurry in psciculture helps in the reduction of cost that would otherwise be invested in supplementary fish feed. This practice has been adopted by some farmers in West Bengal.

2. Mushroom production—Use of BgM has been found profitable in development of excellent media for the culture of edible fungi. This improves the quality and yield of mushrooms. The growth time was 3-4 days faster and pick up time reduced by 3-7 days with increase in yields from 19-26 %. On an average, 100 kg rice stalk can harvest 50-100 kg mushroom. The residues of mushroom and stalk can be used as feed for pig, cattle and other animals as it contains more protein.

Mode of application of BgM—BgM can be applied as needed depending on the species grown/compost preparation and its availability.

Daily—Biogas manure can be applied daily in kitchen gardens, for vegetable and horticultural plants. This functions as a soil conditioner. The biogas manure can be directly applied into an irrigational channel or applied while watering the plants. However, an exhaustive study to evolve a proper system to meet the requirement of different crops based on the output and fertility of soil is necessary.

Fortnightly/Weekly—Storage capacity, availability of biomass for compost making and alternative methods for use of biogas manure through pelletization are to be considered for fortnightly/ weekly applications. For dry land and cash crops, the slurry can be applied weekly/fortnightly.

Monthly—For paddy and horticultural plants BgM can be used periodically where the application of inorganic fertilizer is less as it is found to work as better substitute without any detrimental effect on the crop yield.

CONCLUSION

BgM an effluent of biogas plants has good nutritive value to both plant and soil and should be utilized for increasing crop production as well as soil fertility. Presently, various methods of handling and utilization of biogas manure are available. It is essential to maximize the use of nutrients available in the BgM for crop production at appropriate time periods as an effective substitute to inorganic fertilizers.

REFERENCES

Anonymous 2008-09. Economic Survey, Government of India, New Delhi.

Anonymous 2009. Akshay Urja, Ministry of New and Renewable Energy, Government of India, New Delhi.

Bhattacharryya, B. C., Hanuman, H. S., Banerjee, R., Bhattacharryya, P. and Singh, R. P. 1993. Practica Hand Book. Indian Institute of Technology, Kharagpur.

Chawla, O. P. 1986. Advances in Biogas Technology. Publications and Information Division, Indian Council of Agricultural Research, New Delhi.

Khandelwal, K. C. and Mahdi, S. S. 1986. Biogas Technology: A Practical Handbook. Tata McGraw-Hill Publishing Company Limited, New Delhi.

Khandelwal, S. K. and Sharma, V. 2004. Dual Use of *Parthenium hysterophorus* following Anaerobic Fermentation: Biogas and Manure. Proceedings of National Seminar on Organic Farming, RCA, MPUAT, Udaipur.

Khandelwal, S. K. and Sharma, V. 2004. Eco-friendly Use of Environmental Hazard – Congress Grass for Biogas Production. Proceedings of the National Seminar on Environment (NSE–13). North-Eastern Hill University, Shillong.

Mathur, A. N. and Rathore, N. S. 1992. Biogas Production, Management and Utilisation. Himanshu Publications, New Delhi.

Mathur, A. N. and Verma, L. N. 1988. Management and Utilisation of Biogas Plant Slurry. Himanshu Publications, New Delhi.

Shyam, M. and Sreenivasa, M. N. 1998. Use of Biogas Plant Spent Slurry in Agriculture. Central Institute of Agricultural Engineering, Bhopal.

Srivastava, P. K., Shukla, B. D. and Ojha, T. P. 1994. Technology and Applications of Biogas. Jain Brothers, New Delhi.

Tiwari, R. C., Sharma, G. S. and Khandelwal, S. K. 2009. Organic Farming for Sustainable Agriculture, Himanshu Publications, New Delhi.

Yawalkar, K. S., Agarwal, J. P. and Bokde 1984. Manures and Fertilizers. Agri-Horticultural Publishing House, Nagpur.

❑❑❑

Chapter **14**

Impact of Organic Cultivation on Soil Health for Long Term Sustainability

S.S. RATHORE* AND KAPILA SHEKHAWAT**

*Senior Scientist, Agronomy, Directorate of Rapeseed-Mustard Research,Bharatpur.

**Scientist, Agronomy, Directorate of Rapeseed-Mustard Research,Bharatpur.

INTRODUCTION

Organic farming is accounting for over 24 million hectares worldwide has been increasing at annual rate of 20% in the last decade (Lotter, 2003, Willer and Yussefi, 2004), and has become a main practice for some crops (Anon, 2004). Organic farming is speedily increasing worldwide. Soil organic matter (SOM) plays a dominant role in soil fertility and crop productivity. Plant growth in organic systems greatly depends on the functions performed by soil microbes, particularly in nutrient supply. Soil is a living, dynamic agri-ecosystem. Healthy soil maintains with microscopic and larger organisms that perform many fundamental functions including converting dead and decaying matter as well as minerals to plant nutrients. Organic matter in soil is one of the most deciding factors in maintaining optimum soil health on long term basis, dead or live soils are defined on the basis of soil organic matter content. The humus provides stability even in most unhealthy soil like heavy clay or sandy soils. Organic matter contains carbon compounds that were formed by living organisms. It covers a wide range of material like crop residues, moss, algae, lichens any parts of animals, manure, droppings, sewage sludge, sawdust, insects, earthworms and microbes. Lack of good soil testing facilities and neglect of organic manures led to degradation of soil health. The organic matter harbour different kinds of soil organisms in the soil, these include bacteria, fungi, nematodes, earthworms and insects etc. These organisms live on or other soil organisms and perform a number of vital processes in soil. Other organisms are involved in transformation of inorganic molecules. There are three main components of organic matter in soils, dead forms of organic material - mostly dead plant parts, living parts of plants - mostly roots and living soil micro flora and fauna. Organic agriculture builds the health of the soil, providing the foundation for healthy crops and a livelihood for good stewards of the land.

OPTIMUM SOIL HEALTH

Soil is a key natural resource and sound soil health is the integrated effect of management on most soil properties that determine crop productivity and sustainability. In general, soil organic carbon (or organic matter) is considered to be the universal indicator of soil quality (Rasmussen and Collins, 1991). Good soil quality not only produces good crop yield, but also maintains environmental quality and consequently plant, animal and human health.In comparison with chemical intensive farming, organic farming has potential benefits in promoting soil structure formation (Pulleman *et al.*, 2003), enhancing soil biodiversity (Oehl *et al.*, 2004). Soil organic matter is the most fundamental source of fertility in organic agriculture and it is important for producers to understand the basics of organic matter cycling in the soil. Soil organic matter is that portion of the soil that consists of biological residues, from plant to animal to microorganism. Organic residues supply not only readily available nutrient sources but also the building blocks of humus. Humus is the product that is left over after decomposition has ended, and it is extremely important in increasing and maintaining soil fertility. It possesses an overall negative charge, which means that positively-charged nutrients are attracted to it. Humus can be thought of as a bank which holds nutrients and can release them in response to plant or microorganism needs.

Diverse residue sources sustain a microbial community that is efficient and has more diversity. For instance, bacteria are associated with high nitrogen materials, while the abundance of fungi increases in relation to high carbon materials. It is important to note that if the C: N ratio of the soil is too high, nitrogen will be in short supply and will be used up by microorganisms before it is available to the crop. Over time, and with good soil management, the cycling of nutrients through microbial biomass will reach equilibrium. At that point nutrients will be readily available for crop needs.

ORGANIC MATTER STATUS

It is estimated that around 700 mt of agricultural waste is available in the country every year, but most of it is not properly used. This implies a theoretical availability of 5 tonnes of organic manure/hectare arable land/year, which is equivalent to about 100 kg NPK/ha/yr.

The improvement of soil quality through increase in soil organic carbon pool has numerous subsidiary benefits. The available data show that crop yields can be increased by 20-70 kg ha^{-1} for wheat, 10-50 kg ha^{-1} for rice, and 30-300 kg ha^{-1} for maize with every 1 Mg ha^{-1} increase in soil organic carbon pool in the root zone. Adoption of recommended management practices on agricultural lands and degraded soils would enhance soil quality including the available water holding capacity, cation exchange capacity, soil aggregation, and susceptibility to crusting and erosion. Increase in soil organic carbon pool by 1 Mg ha^{-1} y^{-1} can increase food grain production by 32 million Mg y^{-1} in developing countries. This strategy would also off-set fossil fuel emissions at the rate of 0·5 Pg C y^{-1} through carbons sequestration in agricultural soils of developing countries (Lal, 2006). SOM contains approximately 58% C; therefore, a factor of 1.72 can be used to convert OC to SOM. Total organic carbon is expressed as percent C per 100 g of soil.

Table 1 : Different fraction of soil organic matter, their particle size and description

S. No.	Soil Organic Matter Fraction	Particle Size (mm)	Description
1.	plant residues	≥ 2.0	recognizable plant shoots and roots
2.	particulate organic matter	0.06 – 2.0	partially decomposed plant material, hyphae, seeds, etc
3.	soil microbial biomass	variable	living pool of soil organic matter, particularly bacteria and fungi
4.	humus	≤ 0.0053	ultimate stage of decomposition, dominated by stable compounds

Organic matter has usually partitioned into to two major parts, active (labile) and stable. Labile soil C pools are especially important because they are more vulnerable to climatic change and disturbance and play vital roles in nutrient cycling. Active part of organic matter is plant litter, the light fraction the biomass and non humic substances not integrated to mineral constitutes and active pool of organic matter is ready source of nutrients to the plants, where as stable pool of organic matter is reservoir of nutrients and it is important from long term balance of the soil. Active fraction is particularly important in maintaining the soil productivity under sustainable agriculture.

AGRONOMIC PRACTICES FOR INCREASE IN SOIL ORGANIC MATTER

The organic matter management of degraded soils and the adoption of recommended management practices on agricultural soils can reverse degradative trends and lead to SOC sequestration. The global potential of SOC sequestration and restoration of degraded/ desertified soils is estimated at 0.6 to 1.2 Pg C/y for about 50 years with a cumulative sink capacity of 30 to 60 Pg. The SOC sequestration is a cost effective strategy of mitigating the climate change during the first 2 to 3 decades of the 21st century. While improving soil quality, biomass productivity and enhanced environment quality. (Lal, R. 2003).

- Organic mulching directly provides organic C inputs to soil, and has been used to effectively suppress weeds and reduce soil erosion in organic farming systems (Jordan, 2004). It is also effective in conserving soil moisture and buffering drastic changes in soil temperature (Naeini and Cook, 2000), which can be of special importance in sandy soils where large fluctuations of soil moisture and temperature often occur.
- Conservation tillage- novel feature of these tillage systems is the retention of a crop residue mulch just after crop establishment to ensure an adequate soil cover. Crop residues mulching can be defined as a technology whereby at the time of crop emergence at least 30% of the soil surface is covered by organic residue of the previous crop.
- Farm Yard Manures, green manuring, compost, vermicompost and precision farming strategies, conversion of monoculture to complex diverse cropping

systems, meadow-based rotations and winter cover crops, and establishing perennial vegetation on contours and steep slopes.

- Safe disposal of sewage and sludge after treatment
- Cover cropping of short duration leguminous crops during off season

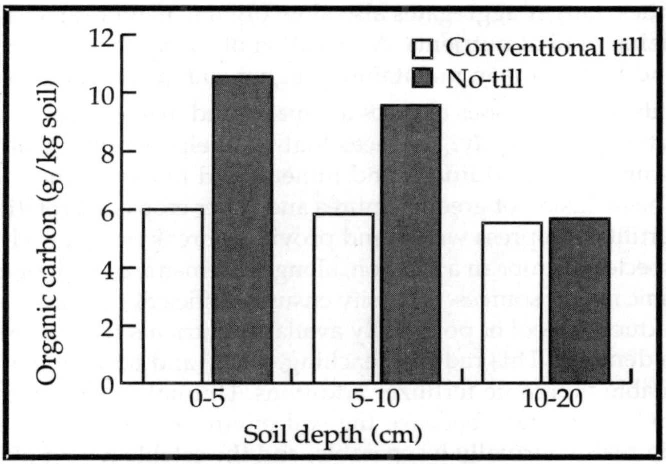

Fig. 1 : Effect of conventional tillage and no-tillage on soil organic carbon

DECREASE IN SOIL ORGANIC MATTER

Most agricultural soils have lost 50 to 70% of their original soil organic carbon (SOC) pool, and the depletion is exacerbated by further soil degradation and desertification (Lal, R. 2003). Other practices led to decline in soil organic carbon

- Excessive tillage operations during field preparation
- Exposing of soil by destroying vegetation cover on the soils, overall, cropping led to less organic matter to the soil than native vegetation.
- Slash and burning agricultural practices
- Excessive dependant on high analysis chemical fertilisers
- Poor soil conservation practices led to loss of fertile top soil
- The wetter a soil is, the less oxygen is available for organic matter to decay, so that it build up
- Organic matter decays more rapidly at higher temperatures, so soils in warmer climates tend to contain less organic matter than those in cooler climates.
- Clay soils tend to have more organic matter than coarse soils. Coarse soils are better aerated, and the presence of oxygen results in a more rapid decay of organic matter.

EFFECT OF ORGANIC MATTER ON SOIL HEALTH

Soil fertility is the combined effects of chemical, physical and biological characteristics of the soil and organic matter has profound positive effect on these soil properties, thus increasing overall soil fertility. The maintenance of soil organic matter (SOM) is desirable for long-term sustainability because of the manifold positive effects of organic matter on nutrient status, water holding capacity and physical structure. High organic matter content also has a positive effect on soil physical properties. For example, soils with high organic matter content contain a greater abundance of water-stable aggregates and have a greater exchange capacity, which translates into better structure and water-holding and nutrient absorption capacities. Larger aggregates also slow organic matter degradation, producing a slowly mineralizing pool of nutrients. Application of *Gliricidia* loppings was superior to *Sorghum* stover and no residue in maintaining higher soil quality index values.

The benefits of healthy soils to crops are many, and management is the key to ensure that a soil is functioning correctly. Practices that can help to build healthy soils include crop rotation, organic matter additions and mineral and nutrient amendments to correct soil imbalances. The inclusion of green manures and cover crops in a rotation is an excellent way to sponsor fertility, suppress weeds and provide a break in pest cycles. Incorporating several different species of crops in a rotation, along with manures and/or compost, ensures a diversity of organic matter sources. Diversity ensures sufficient organic C and N for humus formation and produces a pool of potentially available nutrients that can become mobilized according to crop demand. This reduces leaching, waste and toxicity that can result from immediately-available inorganic fertilizer additions. Ultimately, managing for good soil fertility is extremely important because the soil environment and the surrounding air environment are in reality virtually inseparable, and the establishment of a functional and stable system in one environment can have far-reaching impacts in the other.

PHYSICAL PROPERTIES

At the end of cropping season, addition of organic materials had increased organic carbon, aggregate stability, moisture retention capacity and infiltration rate of the surface soil while reducing the bulk density. Application of inorganic fertilizer alone decreased the stability of macro aggregates and moisture retention capacity but increased the bulk density values (Sarkar *et al.*, 2003)

Soil physical conditions are defined by using the terms bulk density, penetrability, water infiltration rate, water holding capacity, and erodibility. Organic farming practices resulted in good physical conditions of the soils, which results in

- Plant roots can grow through the soil without restriction.
- Air, water, and nutrients needed by plants and soil organisms can move through the soil with relative ease.
- Water from rainfall or irrigation seeps into the soil, rather than flowing over the soil surface as runoff.
- The decrease in BD as a result of organic matter application is due to a dilution effect resulting from the mixing of the added organic matter (OM) with the more dense mineral fraction of the soil.
- Because of increased porosity with organic matter additions, the saturated hydraulic conductivity (HC) is expected to increase.

Organic matter plays an important role in nutrient availability and soil aggregate stability. Soil productivity decreases when soil organic matter (SOM) declines (Bauer and Black, 1994). More water-stable aggregates (>250 μm) were found in a silt loam soil after continuous wheat (*Triticum aestivum* L.) as compared with a wheat–fallow rotation system because of continuous addition of residue (Monreal *et al.*, 1995). After 10 yr of no-tillage continuous corn, Karlen *et al.* (1994), found that removal, maintenance, and doubling of crop residue resulted in 42, 46, and 60% wet aggregate stability, respectively. Water-stable aggregates are divided into macro- and microaggregates, and different organic matter fractions are responsible for the formation of these aggregates. Temporary and transient binding agents are thought to be for macroaggregation (>250 μm) (Tisdall and Oades, 1982). Persistent binding agents are important in microaggregation (<250 μm) of soil (Tisdall and Oades, 1982). Macroaggregates are more susceptible to physical disruption because of the labile nature of binding agents.

CHEMICAL PROPERTIES

The soil chemical properties are explained in terms of soil fertility, salinity, and acidity or alkalinity. Soil fertility can be further divided into nutrient availability for plant uptake, the soil's nutrient-holding capacity, and the balances among plant-available nutrients. Organic matter in the soil creates following chemical properties in the soil for optimum soil health which led to high crop productivity.

- Soils have a near-neutral pH.
- Sufficient nutrients are available for better crop growth, but not in excess that causes plant toxicities or contaminate nearby soil or water sources.
- Soil nutrients are in forms available for plant uptake, but they are held sufficiently enough not to be easily leached or carried away by runoff.
- The availability of plant nutrients is sufficiently balanced to promote healthy plant growth and microbial activity.
- Soils do not contain toxins or heavy metals in concentrations high enough to inhibit plant growth or the growth of beneficial soil organisms.
- Soils contain enough moisture to facilitate nutrient movement to plant roots.
- Soils contain sufficient oxygen for plant growth and the growth of soil organisms.
- Soils contain relatively high levels of organic matter, which helps them hold water and nutrients.
- The application of FYM significantly increased the saturated hydraulic conductivity and decreased the electrical conductivity and exchangeable sodium percentage (ESP).
- Organic matter makes soil resilience to chemical reaction, thus provide stability of soil health.
- Improved fertilizer efficiency;
- Longlife N – for example, urea performs 60–80 days longer;
- Improved nutrient uptake, particularly of P and Ca;
- Salinity management – humates "buffer" plants from excess sodium;
- Organic humates are a catalyst for increasing soil C levels.

Organic matter has both a direct and indirect effect on the availability of nutrients for plant growth. In addition to serving as a source of N, P, S through its mineralization by soil microorganisms, organic matter influences the supply of nutrients from other sources (for example, organic matter is required as an energy source for N-fixing bacteria). A factor that needs to be taken into consideration in evaluating humus as a source of nutrient is the cropping history. When soils are first placed under cultivation, the humus content generally declines over a period of 10 to 30 years until a new equilibrium level is attained. At equilibrium, any nutrients liberated by microbial activity must be compensated for by incorporation of equal amounts into newly formed humus. Some compounds, such as certain phenolic acids, have phytotoxic properties; others, such as the auxins, enhance plant growth.

BIOLOGICAL PROPERTIES

The organic matter serves as a source of energy for both macro- and microfaunal organisms in the soil. The soil bacteria, actinomycetes and fungi are related to humus content. Earthworms and other faunal organisms are strongly affected by the quantity of plant residue material returned to the soil. Plant growth in organic systems greatly depends on the functions performed by soil microbes; particularly in nutrient supply as only nutrient sources is the organic matter. Improving soil physical and chemical properties is important for both conventional and organic production, but improving biological properties is particularly important for organic production. Soil scientists measure soil biological health in terms of microbial biomass, microbial communities, and rate of organic matter decomposition. Organic substances in soil can have a direct physiological effect on plant growth.

Organic farming system relies on nutrients released through the decomposition of plant and animal residues. Decomposition is a biological process involving a variety of soil organisms, including insects, worms, nematodes, fungi, bacteria, and algae etc. Soil organic matter improves biological properties of the soil

- Plant and animal residues added to the soil are readily broken down and decomposed so that plant nutrients become available.
- A good soil structure, provided by stable organic compounds, remains following the decomposition of plant and animal materials.
- Soil is well-aggregated; that is, it is composed of soft clumps held together with fungal threads and bacterial gel.
- Legumes form healthy nodules and fix abundant nitrogen, especially in nitrogen-depleted soils.
- Plants have a relatively high resistance to soil-borne diseases.

Organic matter in soil; harbours various types of living soil flora and fauna, which improves soil fertility on long term basis.

SN.	Organisms involved	Functions
1.	Bioturbating invertebrates and plant roots, mycorrhizae and some other micro-organisms	Maintenance of soil structure
2.	Most bioturbating invertebrates and plant roots	Regulation of soil hydrological processes

contd. ...

... contd.

SN.	Organisms involved	Functions
3.	Mostly micro-organisms and plant roots, some C protected in large compact biogenic invertebrate aggregates	Gas exchange and carbon sequestration (accumulation in soil)
4.	Mostly micro-organisms	Soil detoxification
5.	Mostly micro-organisms and plant roots, some soil- and litter-feeding invertebrates	Nutrient cycling
6.	Various saprophytic and litter-feeding invertebrates (detritivores), fungi, bacteria, actinomycetes and other micro-organisms	Decomposition of organic matter
7.	Rhizobia, mycorrhizae, actinomycetes, diazotrophic bacteria and various other rhizosphere micro-organisms, ants	Symbiotic and asymbiotic relationships with plants and their roots

It is well known fact that the earthworms are an important indicator of soil fertility (Christensen, 1988).The earthworm biomass, density and diversity were found to be highest on the organic soils compared with the conventional fertilizer fed soils (Pfiffner and Mäder, 1997). Several studies have shown that a relationship between earthworms and soil properties or land-use, respectively, exists. It is widely known that many of the factors influencing the incidense of pathogenic organisms in soil are directly or indirectly influenced by organic matter. For example, a plentiful supply of organic matter may favor the growth of saprophytic organisms relative to parasitic ones and thereby reduce populations of the latter. Biologically active compounds in soil, such as antibiotics and certain phenolic acids, may enhance the ability of certain plants to resist attack by pathogens.

MINIMIZE SOIL EROSION

Soil erosion is the loss of surface soil to forces of wind and water. Soil erosion reduces soil productivity and can cause water and air pollution. Minimizing soil erosion is critical for sustainable agricultural production, because the top layer of soil has properties—physical, chemical, and biological—that are much more favorable for crop production than the lower layers. The topsoil also contains more organic matter than the lower layers. This native soil organic matter—which represents centuries of plant growth, death, and decomposition in the soil—can both protect soil against erosion and be easily lost to it. Soil aggregates, held together by organic compounds, facilitate water infiltration and enhance water-holding capacity. They also are slow to erode because of their size and weight. However, if soil aggregates are broken down—by dry conditions, excessive tillage, or by raindrop impact—the resulting organic matter particles are very lightweight and easily removed by erosion.

Tillage prepares land for seeding or transplanting. Cultivation practices—implemented either before or after crops are in the ground—manage weeds and improve soil aeration and water infiltration. Traditional clean tillage disrupts soil organisms, reducing their numbers and, often, their diversity. Clean tillage also breaks down soil aggregates and produces a bare soil that is vulnerable to erosion and more subject to temperature changes than soils that have a mulch or vegetation cover. Under direct exposure to sun and high temperatures, soil organisms break down soil organic matter rapidly. The rate of organic matter decomposition further increases when tillage breaks soil aggregates into smaller pieces, increasing their surface area and exposure to oxygen. While a rapid breakdown of organic matter can benefit plants by increasing nutrient availability, a substantial decrease in organic matter reduces aggregate stability, decreases soil tilth, and increases the potential for nutrient loss through leaching and runoff.

OTHER EFFECTS

The long term studies indicates that, the organic farming system was more effective than the conventional farming system in reducing soil erosion and, therefore, in maintaining soil productivity. The organic matter minimises the effect of drought on soils and on the crops. The poor utilization of rainfall is more due to inadequate land management practices (i.e. burning of crop residues, excessive tillage, eliminating hedges, etc.) than because of natural phenomena (relief, slope, rainfall intensity) that reduce organic matter levels, destroy soil structure, eliminate beneficial soil fauna and do not favour water infiltration. To minimize the impact of drought, soil needs to capture the rainwater that falls on it, store as much of that water as possible for future plant use, and allow for plant roots to penetrate and proliferate. The capacity of soil to retain and release water depends on a broad range of factors such as soil texture, soil depth, soil architecture (physical structure including pores), organic matter content and biological activity. The addition of organic matter to the soil usually increases the water holding capacity of the soil. This is because the addition of organic matter augments the number of micropores and macropores in the soil. Certain types of soil organic matter can hold up to 20 times their weight in water.

Table 2 : Soil organic properties and their impact on soil quality

S.No.	Soil quality	Particular	Impact on the soil
1.	Colour	Organic matter caused dark colour of the soil	Absorb more solar radiation
2.	Water retention	OM can retain more than 20 times of its weight in water	This improve soil water holding capacity
3.	With clay material	Make stable soil aggregates	Stabilises soil structures, better aeration
4.	Buffer action	OM exhibits pH buffering	Maintains uniform reaction in the soil
5.	Cation exchange	The CEC of humus range from 350- 1400 meq/100 gm of soil	High CEC, means better availability of nutrients to crop
6.	Mineralization	Decomposition of OM resulted CO_2, NH_4, NO_3, H_2PO_4 and SO_4	Increase plant nutrient availability

contd. ...

... contd.

S.No.	Soil quality	Particular	Impact on the soil
7.	Solubility in water	Because of association of OM with clay, it is insoluble	Meagre amount of OM lost by leaching
8.	Chelation	Form stable complexes with element and cations	It may enhance nutrient availability to the plants

CONCLUSION

Organic matter affects physical, chemical and biological properties of the soil and its overall health. The maintenance of soil organic matter levels and the optimization of nutrient recycling are essential to the sustained productivity of agricultural systems. Maintaining soil organic matter content requires a balance between addition and decomposition rates. Conventional, intensive agricultural practices have significantly increased crop productivity. But, serious issue is being raised about the energy-intensive nature of these systems and their adverse effects on soil productivity and environmental quality. This concern has led to an increasing interest in organic farming systems because they may reduce some of the negative effects of conventional agriculture on the environment. The long-term effects of organic and chemical intensive farming on soil properties showed that the organically-farmed soil had significantly higher organic matter content, thicker topsoil depth, higher polysaccharide content, lower modulus of rupture and less soil erosion than the conventionally-farmed soil. It is proved beyond doubt that in the long term, the organic farming system was more effective than the conventional farming system in reducing soil erosion and, therefore, in maintaining soil productivity.

REFERENCES

Anon, 2004 . Organic farming enters the mainstream, *Nature* (London) 428, p. 783.

Black A.L. 1994. Quantification of the effect of soil organic matter content on soil productivity. *Soil Science Society of America Journal*;58:185-193.

Christensen, O. 1988. Lumbricide earthworms as bio-indicators relative to soil factors in different agro-ecosystems. In: Veeresh, G.K., Rajagopal, D., Viraktamath, C.A. (Eds.), Advances in Management and Conservation of Soil Fauna, Oxford and IBH Publishing, New Dehli, Bombay, Calcutta, pp. 839–849.

Jordan, C.F. 2004. Organic farming and agroforestry: Alleycropping for mulch production for organic farms of southeastern United States, Agroforestry Systems 61/62, pp. 79–90.

Karlen D.L., Wollenhaupt N.C., Erbach D.C., Swan J.B., Eash N.S., Jordhal J.L. 1994. Long-term tillage effects on soil quality. *Soil Tillage Research*; 32:313-327.

Lal, R. 2006. Enhancing crop yields in the developing countries through restoration of the soil organic carbon pool in agricultural lands. Land Degradation & Development, 17: (2): 197-209.

Lotter, D.W .2003. Organic agriculture, *Journal of Sustainable Agriculture* 21, pp. 59–128.

Monreal C.M., Schnitzer M., Schulten H.R., Campbell C.A., Anderson D.W. 1995. Soil organic structures in macro and microaggregates of a cultivated brown chernozem. *Soil Biology and Biochemistry*; 27:845-853.

Naeini S.A.R.M. and Cook, H.F. 2000 . Influence of municipal compost on temperature, water, nutrient status and the yield of maize in a temperate soil, *Soil Use and Management* 16, pp. 215–221.

Oehl, E. Sieverding, P. Mäder, D. Dubois, K. Ineichen, T. Boller and A. Wiemken. 2004. Impact of long-term conventional and organic farming on the diversity of arbuscular mycorrhizal fungi, *Oecologia* 138 , pp. 574–583.

Pfiffner, L., Mäder, P. 1997. Effects of biodynamic, organic and conventional production systems on earthworm populations. Entomological Research in Organic Agriculture (Special edition of Biological Agriculture and Horticulture), 15, pp. 3–10.

Pulleman, A. Jongmans, J. Marinissen and J. Bouma, 2003. Effects of organic versus conventional arable farming on soil structure and organic matter dynamics in a marine loam in the Netherlands, Soil Use and Management 19, pp. 157–165.

Rasmussen and Collins. 1991. P.E. Rasmussen and H.P. Collins, Long-term impacts of tillage, fertilizer and crop residue on soil organic matter in temperate semi arid regions, *Adv. Agron*. 45, pp. 93–134.

Sarkar, S. Singh, S. R. and Singh, R. P. 2003. The effect of organic and inorganic fertilizers on soil physical condition and the productivity of a rice–lentil cropping sequence in India. *The Journal of Agricultural Science*, 140 , pp 419-425.

Tandon, H. L. S., In Plant Nutrient Needs. 1997. Efficiency and Policy Issues: 2000–2025, *National Academy of Agricultural Sciences*, New Delhi, pp. 15–28.

Tisdall J.M., Oades J.M. 1982. Organic matter and water stable aggregates in soils. *Journal of Soil Science*; 33: 141-163.

Willer and Yussefi. 2004. The World of Organic Agriculture: Statistics and Emerging Trends, International Federation of Organic Agriculture Movements, Bonn, Germany pp. 167.

□□□

Chapter **15**

Sustainable Crop Production through Organic Farming, Its Waste and Management

R.C. Tiwari

Associate Professor, Department of Agronomy, Rajasthan College of Agriculture, Maharana Pratap University of Agriculture and Technology, Udaipur

INTRODUCTION

The traditional agriculture practiced by our forefather farmers was essentially organic. The use of animal excrement on cultivated soils was common as far back as agricultural records can be traced. The use of mineral salts as fertilizer material to promote plant growth was established only during the last 100 years. This was the development over years based on the famous "Mineral Nutrition Theory" propounded in 1840 by German Chemist Justus von Liebig and he had demonstrated the essentiality of potassium in 1845. The necessity for supplying plants with N was not recognized until 1857 when Lawes, Gilbert and Pugh of Rothamsted Experiment Station, England established the essentiality of nitrogen. Similarly, albeit, bones were one of the first fertilizing materials along with animal excrement, guano, fish wastes and wood ash etc. Basic slag as a source of P was first used in England in 1884. This in nut shell traces the origin, development and use of N, P and K as fertilizer elements in the past.

It is estimated that the consumption of chemical fertilizers increased seven- fold, pesticides by 375 times while the food production had just doubled during the first 20 years of the launch of green revolution in India (Palaniappan and Annadurai 1999). Increased use of chemicals under intensive cultivation has disturbed the harmony existing among soil, plant and microbial population (Ghosh, 1999).

There are reports of frustration among farmers cultivating cotton and paddy and ended up in suffering severe losses and even ending their life. Such news items appearing more often from Andhra Pradesh, Maharastra, (Vidharbha), Kerala and Karnataka have been most painful and disheartening to all of us.

FOOD CHAIN

According to Angelina Briones (1999) the food chain begins in farms (agriculture and marine) and ends on the dining table. Basically, everything absorbed by plants from the soil and water becomes part of the food chain. These things include elements and substances from agrochemicals (pesticides, herbicides, fungicides), air pollutants (sulphur dioxide and nitrogen oxides) that cause acid rain; shelf-life extenders, and even elements that become soluble in cooking wares and food containers. The quantity of pesticides consumed through food in different countries is presented in table 1.

Table 1 : Quantity of pesticides consumed through food

Country	Pesticides consumed through (mg^{-1} $person^{-1}$)
USA	7.6
United European Countries	12.0
Canada	13.3
Germany	149.0
Europe	156.0
India – Vegetarians	356.3
India – Non vegetarians	362.5

India stands at the top with the highest quantity of pesticides being consumed. Farm animals treated with antibiotics and other drugs contribute even more to the food chain. For sea food, everything ingested by the fishes, shrimps, dams, oysters etc. from their marine environment becomes part of the food chain. As a consequence, the thrust since last decade has been more towards sustainable farming practices. Even globally there is an ever-increasing consciousness of conservation and protection of all living beings consisting of ecosystem, biodiversity and environment endangered by human interference and indulgence.

WHAT IS ORGANIC FARMING?

Organic farming is enunciated by IFOAM (International Federation of Organic Agriculture Movement) covers all aspects of operating farms that promote environmentally, culturally and economically sound production of food, fiber, herbs and fuel wood. Organic farming is the science and art of producing crops and livestock. It has three distinct features of appropriate cropping patterns (crop rotations, intercropping relay cropping, alley cropping, mosaicking with multipurpose tree species) and livestock integration; use of organic materials (green manure, *Azolla,* crop residues, animal manure, guano, farm wastes) as natural sources and soil conditioners; use of non-chemical methods for control of pests, diseases and weeds (Angelina-Briones 1999). Thus the organic farming may be defined comprehensively as environment friendly ecological production system that promotes and enhances biodiversity, biological activities. It is based on minimum use of off-farm inputs and management practices that restore, maintain and enhance ecological harmony (Narayanan 2005). Precisely organic farming encompasses the following aspects:

- It includes production technology under agriculture, animal husbandry, poultry farming and aquaculture etc.
- It aims to do away with all synthetic or artificial inputs which include chemicals as well as fertilizers plant protection chemicals, weedicides and plant and animal harmones etc.
- The input from seed to seed, calf to calf and egg to egg should only be from natural sources and processes.
- At post harvest stage, food processing should ensure non-use of any chemicals, synthetics and preservatives or colouring or flavouring agents.
- It should have increased use of on-farm inputs and minimum use of off-farm inputs.
- the allowable inputs shall include crop residues, green manure, green leaf manure, farm wastes, FYM, oilcakes, animal and poultry wastes and their excreta, slaughter house wastes, agricultural by-products wastes from sugar mills, coir industries, paper mills, urban and rural human wastes as compost, bird guano, fish wastes etc. The list is large and innumerable. The estimated recyclable wastes from India are presented in table 2.

Table 2 : Recycle wastes available for organic farming in India

S. No.	Name of wastes	Quantity
1.	Urban solid wastes	27.4 Million tones per year
2.	Urban liquid wastes	12145 Million liters per day
3.	Sugar mill wastes (Press mud)	9 Million tones per year
4.	Wastes from food and food processing industries	4.5 Million tones per year
5.	Dairy wastes	60 million liters per day
6.	Other agricultural wastes	350 Million tones per year

Source: Selvaraj *et al.* (2005)

Principal Aims of Organic Farming as Embodied in the IFOAM

- To produce food of high nutritional quality in sufficient quantities
- To interact in a constructive and life-enhancing way with natural systems and cycles
- To encourage and enhance biological cycles within the farming system, increasing micro-organisms, soil fauna and flora, plants and animals
- To maintain and increase long-term fertility of soils
- To prepare the healthy use and proper care of water, water resources and all life therein
- To help in the conservation of soil and water
- To use, as far as possible, renewable resources in the locally-organized agricultural systems
- To work, as far as possible, with materials and substances which can be reused or recycled, either on the farm or elsewhere

- To give all livestock life conditions which allow them to perform the basic aspects of their innate behavior
- To minimize all forms of pollution that may result from agricultural practices
- To maintain the genetic diversity of the agricultural system and its surroundings including the protection of plant and wildlife habitats
- To allow every one involved in organic production and processing a quality of life that conforms to the Human Rights charter to cover their basic needs and obtain an adequate return and satisfaction from this work, including a safe working environment
- To consider the wider social and ecological impact of the farming system
- To produce nonfood products out of renewable recourses, which are fully biodegradable.

Bioproducts in Organic Farming

Bioproducts such as biocontrol agents and biofertilizers are ecofriendly, hazardless, effective and economic inputs in organic farming. The recognized biocontrol agents and biofertilizers as listed by Kannaiyan *et al.* (2005) are given below:

Biocontrol Agents

Microbial pesticide: These are based on the beneficial and ecofriendly micro-organisms of fungal, bacterial, viral and protozoan origin. Generally they are target specific, not only control pests but also promote growth of crops (*e.g. Pseudomonas*)

1. *incorporated protectants:* These are based on pesticide substances that the plant develop which cause resistance to specific pest from introduced genetic material (Bt protein)
2. *Botanical:* They are naturally occurring substances that control pests by non-toxic mechanism (*e.g.* Neem products)
3. *Natural enemies:* There are several predators and parasites occurring in nature, which keep the pest population check (*e.g. Trichogramma chilonis*)
4. *Sex attractants:* These are biochemical's (pheromones) extracted from bodies of insects which attract opposite sex.

Biofertilizers

There are two groups: N fixing organisms such as *Rhizobium, Azospirillum, Azotobacter, Azolla*, BGA and phosphorus solubilizing organisms like phosphobacteria and *Mycorrihiza.*

The dung/excreta production, nutrient content in excreta of animals and the oil cakes is presented in table 3, 4, 5 and 6.

Table 3 : Dung/ excreta production per animal year^{-1} (dry weight basis)

Livestock	Range (t/year)	Average (t/year)
Cattle	0.4-1.8	1.10
Buffalo	0.8-1.9	1.35

contd. ...

... contd.

Livestock	Range (t/year)	Average (t/year)
Camel	0.7-0.9	0.80
Horse and donkey	0.4-0.6	0.50
Pig	0.2-0.3	0.25
Sheep and goat	0.1-0.2	0.15
Hen	0.14	0.14

Source: Tiwari *et al*. (2009)

Table 4 : Nutrient content in excreta of different animals

Material	Nutrient (% on dry basis)				
	N	P_2O_5	K_2O	Ca	Mg
Pig dung	2.27	3.1	1.8	0.21	0.54
Cow dung	1.74	1.7	0.60	0.37	0.53
Horse dung	1.07	2.1	3.6	0.26	0.49
Camel dung	1.51	0.35	1.8	0.70	0.69
Poultry excreta	2.17	2.0	4.2	2.28	1.39
Goat and sheep excreta	0.65	0.50	0.03	-	-
	Micronutrients				
Materials	**Fe**	**Zn**	**Mn**	**Cu**	**B**
Pig	1200	50	70	8.9	-
Poultry excreta	1400	90	210	7.1	5.0
Goat and sheep excreta	-	2570	150	61	4600

Tiwari *et al*. (2009)

Table 5 : Nutrients from the excreta of animals (per year per animal)

S. No.	Animal/ excreta	Nutrient content of excreta/ year/ animal (kg)			Nutrient content of excreta/ year/ animal- Its equivalent to fertilizers (kg)		
		N	P	K	Urea	Super phosphate	Murate of Potash
1	Cattle (dung + urine)	42	10.5	31.5	91	66	52
2	Goat + sheep (dung + urine)	1.76	0.65	1.85	4	4	3
3	Poultry manure	0.23	0.15	0.08	0.5	1	0.14
4	Horse (dung + urine)	21	7.5	16.5	46	47	27

Source: Selvaraj *et al*. (2005)

Table 6 : Nutrient content of oil cakes

Oil cakes	Nitrogen	Phosphorus	Potassium
Non – edible			
1. Castor	5.5	1.8	1.0
2. Mahua	2.5	0.9	1.8
3. Neem	5.2	1.0	1.4
4. Sunflower	4.8	1.4	1.2
5. Pongamia	3.9	0.9	1.3
Edible			
1. Coconut	3.0	1.8	1.3
2. Cotton	6.4	2.8	2.1
3. Groundnut	7.0	1.5	1.3
4. Sesame	6.2	2.0	1.2
5. Niger	4.7	1.8	1.3
6. Rape seed	5.1	1.8	1.2

Source: Selvaraj *et al.* (2005)

Since it is not possible to recycle all the available wastes, part of it can be diverted to supplement the increasing demand of nutrients and earthworms can help in improving the recycling process known as Vermicompost.

EARTHWORM FOR COMPOSTING

India has about 3000 species of earthworms, which are adapted to a range of environment and vermiculture needs. Earthworms can be divided into the following two broad categories:

1. Epigamic: the surface living earthworms
2. Epianecic: the burrowing worms

Epigamic or manure worms are found on the surface and are reddish brown in colour. They do not process the soil but are efficient in composting of organic wastes and enhance the bio- degradation. When maintained in captivity under semi- natural conditions, they remain active through out the year.

Epianecic species burrow and mix the soil from different horizons in the profile. They ingest organic and soil fractions and bring organic- minerals crumbs to surface.

REPRODUCTION OF EARTHWORMS

The reproduction of earthworms depends on temperature, moisture and availability of food materials. Under optimum conditions earthworm becomes adults in 4 weeks and reproduces. One adult gives 2- 3 cocoons in a week till the density of worms restricts the rate of reproduction. Optimum temperature range is 25- 30° C and moisture is about 80 per cent. Life cycle of earth worm is presented in fig. 1.

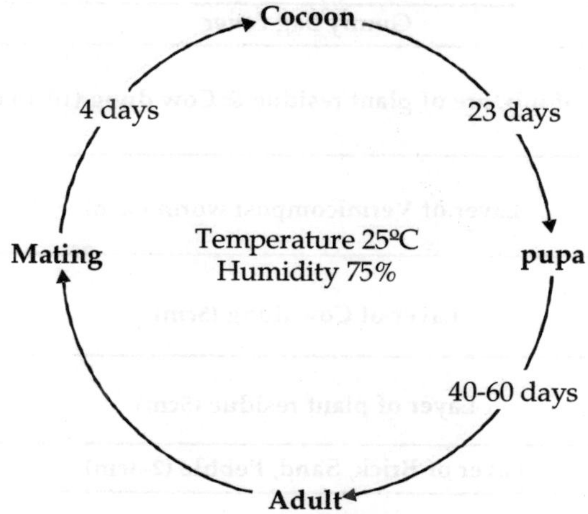

Fig. 1 : Life cycle of earthworm

Comparative production of conventional compost and Vermicompost

Pit of 120 cuft. Gives about one ten of compost per year by conventional method, whereas, 8- 10 tones of compost is produced when it is medicated by earthworm. When sole cow dung is used and 4- 5 tones with dung and residue.

Preparation of compost

Vermicomposting requires temperature in the range of 25- 30° C, optimum moisture and dark conditions. Temperature can be maintained by providing shadow with trees, sheds etc. it checks direct sunlight especially in summer periods. Moisture can be maintained by sprinkling water as and when required especially during summers and protect the beds from rains to have excess waters. Dark conditions are maintained by covering beds with gunny bags, tripal etc.

- **Vermicomposting in shallow beds:**

  Prepare a bed of 40- 50′ length, 3′ width and 1′ depth and cover it with shed of 4- 4.5′ height to protect the bed from direct sunlight and rain.

  At the bottom, provides 2-3 cm thick layer of broken pieces, pebbles etc. this layer is inhabited by earth- worms. Over this layer, prepare a 50 cm thick layer of partially degraded cow dung. Moist the layer by sprinkling water. 10- 15 cm thick layer of mixture of grasses, leaves, crop residues and dung follows Vermicompost layer. Cover the pit with gunny bags and sprinkling water regularly as required to maintain the moisture (about 30%). This layering of compost bed is shown in fig. 2

  Earthworm start ingesting decaying lignocelluloses materials and leaves, branches, roots and other plant residue are converted in to compost in about 2 months periods. Worms leave their casting over the surface till composting is completed and they come to the surface. Change of colour if mixture to black and presence of worms on surface indicate that compost is ready. Bed of 40 X 3 X 1 cuft. gives about 1 ton of compost is ready. Subsequently one can harvest one ton of compost every month.

Gunny bag cover
Layer of mixture of plant residue & Cow dung (10-15 cm)
Layer of Vermicompost worm (5cm)
Layer of Cow dung (5cm)
Layer of plant residue (5cm)
Layer of Brick, Sand, Pebble (2-3cm)

Fig. 2 : Layering of compost bed

- **Vermicomposting in pit:**
  Prepare a pit of 180 cm length, 90 cm width and 90 cm height at elevated places and under shadow of trees etc. At the bottom prepare a layer (15- 10 cm thick) of brick pieces, pebbles and sand keeping raised from center. Over this prepare a layer of organic residue (5.0 cm) followed by 5.0 thick from the layer of partially decomposed cow dung or compost. This layer is over layered by a 5.0 cm thick layer of Vermicompost containing live earthworms 200-300, and cocoons. After about four weeks, another layer of 5.0 cm thick from the mixture cow dung, leaves, grasses etc. is prepared and moist the layers. This layer is repeated every 4- 5 days till the pit is filled completely. Always keep the pit covered with gunny bags and moist by sprinkling water as and when required. When the compost is ready stop watering to dry the surface layers so that worms migrate to deeper layers and harvest surface layers. Since than use the compost. One pit gives about 1 ton of compost every month. This type of layering of compost pit is shown in fig. 3

Gunny bag cover
Plant residue & Cow dung V(5 cm)
Plant residue & Cow dung IV(5 cm)
Plant residue & Cow dung III(5 cm)
Plant residue & Cow dung II(5 cm)
Plant residue & Cow dung I(5 cm
Vermicompost worm (5cm)
Cow dung (5cm)
Plant residue (5cm)
Layer of Brick, Sand, Pebble (5-10cm)

Fig. 3 : Layering of compost pit

- **Vermicompost in box:**

  Composting in box is done where it is required in small scale for kitchen garden etc. Take a box of about 60 cm X 50 cm X 50 cm. at the bottom put a layer of 5 cm thick from stone degrading residues of corn, sorghum, sugarcane etc. followed by 5- 8 cm thick layer of cow dung. Over the cow dung put a thin layer of fresh Vermicompost containing live adult earthworms- 200 Nos. This layer is over layered by 15- 20 cm thick layer of mixtures containing cow dung, kitchen wastes, garden wastes etc. Cover the surface with jute cloth and moist from time to time to maintain moisture of about of about 30 per cent. Protect the box from hot and cold weather. Compost will be ready in 30- 45 days and harvest by drying surface layers. This layering of organic waste in box is shown in fig. 4.

Gunny bag
Cow dung with kitchen waste & garden waste (15-20cm)
Vermicompost worm (5cm)
Cow dung (5-8cm)
Slow degrading plant residue(5cm)

Fig. 4 : Layering in box

Advantages of Vermicompost

- Vermicompost is a source of soil carbon and humus and activities soil microorganisms involved in transformation of nutrients. It also improves soil structure, drainage, aeration, water holding capacity, cation exchange capacity etc.
- It provides balanced nutrition to crops giving good yields and better quality of products.
- It helps in reducing the number of irrigation and quantity of water for each irrigation.
- It provides resistance to plants pathogens and insect thereby reduces the requirement of agrochemicals.
- It helps in reducing the pollution of soil, water and food products by recycling of farm waste and reduced uses of chemical fertilizers.
- Compared to farmyard manure, Vermicompost is rich in nitrogen, phosphorus.
- It is also a good source of nutrients like zinc, copper, calcium, magnesium, sulfur, cobalt etc.

Table 7 : Acceptable composition of commercial vermicompost

Characteristics	Range
C: N ratio	>2
Ash	<20

contd. ...

... contd.

Characteristics	Range
Moisture	10- 20
P (%)	0.15- 1.5
Colour	Brown black
Odor	Earthy smell
WNC (%)	150- 200
CEC (coml./ kg)	75- 100
As (ppm)	<10
Cd (ppm)	<3
Cr (ppm)	<50
Pb (ppm)	<150
Hg (ppm)	<0.15
Ni (ppm)	< 60
Zn (ppm)	<500
Plastic	<1

PRODUCTIVITY LEVELS ACHIEVED UNDER ORGANIC FARMING

In a study conducted at PAU on maize-potato-onion cropping system (Table 8) that in the first year the productivity levels were low and there were consistent improvement in crop yields with passage of time. On completion of two years of this system, maize crop with chemical treatments gave maximum productivity of 50.8 q/ha which was closely followed by organic treatment where soybean was grown as intercrop in maize (50.6 q/ha) and 49.2 q/ha in integrated approach where 50 %NPK was applied from organic sources and 50 % from chemical fertilizers. The other organic manure treatments given in table produced a reasonably good yield of maize grain in the range 42.2 to 45.8 q/ha. In potato maximum yield (168 q/ha) was obtained in organic treatments receiving one- third each from FYM, Vermi compost and non-edible cake + bio fertilizers containing N and P carriers. The integrated treatment receiving 50 % recommended NPK + 50% N as FYM produced 150 q/ha potato which was 26.1 % more than chemical treatment (119 q/ha). The potato yield in rest of the organic treatments varies from 133-149 q/ha. Similar trend was observed in onion where organic fanning showed discernible edge over chemical treatments. The treatment comprising intercropping of soybean in maize, radish in potato and coriander in onion, gave maximum productivity (231.9 q/ha) (Anonymous 2004). The maize equivalent yield in all the organic treatments varied from 210 to 231.9 q/ha except in one treatment where 50 % N was applied through organic source + bio fertilizer+ rock phosphate which produced maize equivalent yield of 175 q/ha and it was attributed to an undernourished crop.

Table 8 : Maize equivalent yield of maize-potato-onion system grown under organic farming

Treatment	Yield (q/ha)			Maize equivalent yield (q/ha)	Organic carbon (%) after 2 years
	Maize	Potato	Onion		
50% recommended NPK+ 50% N as FYM	49.2	150	166	208.8	0.44
1/3rd N FYM + 1/3rd NVC +1/3rd N Non -edible cake	45.8	160	189	223.2	0.49
1/3rd N FYM + 1/3rd NVC +1/3rd N Non-edible cake+ intercropping	50.6 (3.6 Soybean)	115 (130 Radish)	172 (1.44 green coriander+ seed)	231.9	0.47
+1/3rd N Non-edible cake (no chemical pesticide and herbicide and through agronomic practices)	43.6	149	180	211.2	0.50
50% N as FYM+Bio-fertilizer + rock phosphate +PSB	42.2	107	150	175.0	0.42
1/3rd N FYM + 1/3rd NVC +1/3rd N Non-edible cake+ Bio-fertilizer containing N and P carries	44	168	19.5	228.6	0.49
100% NPK + secondary and micronutrients	50.8	119	158	193.2	0.38
1/3rd N FYM + 1/3rd NVC +1/3rd N Non-edible cake (Dummy plot)	45.2	133	186	210.0	0.47

Table 9 : Rice equivalent yield (t/ha) of different cropping systems under various management practices (Average of two years)

Cropping system	Productivity (t/ha)		
	Chemical	Organic	Integrated
Basmati rice-wheat GM	12.6	13.0	13.6
Turmeric-onion	19.2	36.9	36.6
Summer groundnut-garlic	25.3	29.1	29.4
Maize-Durum wheat-cowpea(f)	11.4	12.6	12.3
Rice-garlic + menthe	24.9	31.0	32.2

The Rice equivalent yield of different cropping systems given in Table 9 clearly elucidated the scope of organic farming under Punjab conditions. The different systems on the completion of two year gave better equivalent yields than chemical farming. However, the integrated approach proved highly appreciable and gave maximum productivity. The rice equivalent yield of different cropping systems in organic, chemical and integrated varied from 12.6-36.9, 11.4-25.3 and 12.3-36.6 t/ha, respectively in different cropping systems (Anonymous 2004).

Rajendran *et al.* (2000) reported cotton yields and economics of organic farming vis-à-vis conventional farming and found that from the 4th year onward returns under organic farming were higher than the conventional farming which continued to rise with lapse of time (Table 10).

Table 10 : Yields and economics of organic farming vis-à-vis conventional farming

Year	Status	Yield q ha^{-1}	Gross income	Premium (20%)	Total (Rs)	Net income (Rs)	Surplus/ deficit over conven-tional
Conventional	-	10	20000	0	20000	9000	0
First year	Year of conversion	5	10000	0	10000	750	-8250
Second year	Year of conversion	5.75	11250	0	11250	3750	-5250
Third year	Organic	6.25	12500	2500	15000	7000	-1500
Fourth year	Organic	7.50	15000	3000	18000	10500	1500
Fifth year	Organic	8.75	17500	3500	21000	13500	4500
Sixty year	Organic	10	20000	4000	24000	1665	7500

It is obvious from the results discussed above that productivity under organic farming may be low during initial conversion years but increases with the lapse of time and reaches higher or at least the level obtained under conventional and integrated nutrient management system of farming.

SOIL HEALTH UNDER ORGANIC FARMING

Reganold *et al.* (1993) reported considerable improvement in soil health under organic farming (Table 11). These are mean values of aggregated soils data from 16 pairs of farms each with organic (bio) and conventional (con) farming.

Table 11 : Mean values of aggregated soils data from 16 pairs of farms each with organic (bio) and conventional (con) farming

Soil property	All "bio" farm	All "Con" farms
Soil property	1.07	1.15*
Bulk density (Mg m^3)	2.84	3.18*

contd. ...

... contd.

Soil property	All "bio" farm	All "Con" farms
Penetration resistance (0 to 20 cm) (Mpa)	4.84*	4.27
Carbon (%)	73.7*	55.4
Respiration (µl O_2 h^{-1} g^{-1})	140.0*	105.9
Mineralizable N (mg kg^{-1})	2.99*	2.59
Ratio of mineralisable N to C (mg g^{-1})	21.5*	19.6
CEC (cmol kg^{-1})**		

* = Significantly different at p<0.01** = Cation exchange capacity in centimoles of cation charge (=) per kilogram of soil

MARKET POTENTIAL FOR ORGANIC FOODS IN INDIA

Menon (2007) conducted a survey of over 3600 consumers in four regions of the country. The market potential estimate for 8 metro cities is given in Table 12 . The Table 13 arranges 20 different food categories in the order of preference.

Table 12 : Potential estimated * for organic foods in top 8 metros in India (at retail level 2005-06 prices – considering organic premium 10-20%)

Study	In rupees Million (Figures rounded off)	
Markets	Accessible potential**	Market potential
Chennai	840	1500
Hyderabad	330	740
Bangalore	690	1280
Mumbai	1480	5680
Ahmedabad	300	780
Pune	280	580
Delhi	1120	2230
Calcutta	570	1540
Top 8 metros	5620	14520

* Based on our primary survey **Through Modern retail format

Table 13 : Market potential for different organic food products in top 8 metros (At retails level 2005-06 prices – Considering Organic Premium 10-20%)

Study products	Accessible potential**		Market potential	
	Rs Million	%	Rs. Million	%
Vegetables	1030	18	3220	22
Fruits	710	13	2460	17
Milk	520	9	1660	11
Dairy product	500	9	1110	8

contd. ...

... contd.

Study products	Accessible potential**		Market potential	
	Rs Million	%	Rs. Million	%
Bakery products	480	9	1860	13
Oils	320	6	590	4
Rice	2	5	460	3
Ready to eat	260	5	360	2
Wheat -Atta	250	5	4700	3
Snacks	220	4	560	4
Frozen foods	220	4	300	2
Dals	180	3	320	2
Health Drinks	170	3	340	2
Canned foods	170	3	230	2
Tea	120	2	230	2
Coffee	100	2	170	1
Condiments	50	1	120	1
Spices	40	1	80	1
Sugar	2.8	0	4.8	0
Baby Food	0.1	0	0.30	0
TOTAL	**5620**	**100**	**14520**	**100**

It is apparent from the above estimates that preference of consumer for organic foods is on the rise and with the growing consciousness among consumers the demand for organic foods will not be limited to metro cities only but will increase in smaller cities and towns also. Furthermore, as the consciousness grows, home-consumers are not likely to rely on organic food certificate by the authorized agencies only but may also buy produce that is certified by the local producers based on producer consumer faith / trust.

AREAS FOR ORGANIC FARMING IN INDIA

As has already been mentioned earlier, productivity levels under organic farming system remains low during the conversion years, it would not be logical to convert commercial agriculture areas of high productivity in the organic farming system especially in view of growing demand for food grains and other agricultural produce. Instead areas under rainfed farming tribal, north east and hilly regions of the country where negligible chemicals are used and thereby organic farming is being practiced by default should be considered for promoting organic farming.

FUTURE THRUST AND CONCLUSION

Organic farming in India is at a nascent stage with a lot of challenges to overcome. Any new technology introduced for agricultural development takes time to spread. All the technologies of the past were not adopted overnight but took considerable time based on benefits. Organic farming has a future in India as we are at the cross-roads concerning the need for reducing the use of synthetic plant protection chemicals.

Considering the various aspects of organic farming presented in the foregoing pages there is an urgent need to address this from the point of view of national food security, nutritional security, maintenance of soil health, environment safety, ecological balance, biodiversity and quality produce. Above all, there is a commitment to leave a good heritage for posterity preserving the mother earth.

In this context there is a greater need to create awareness of organic farming, explore its potentialities discover its complexities, evaluate its behaviour, asses its benefits and learn to adopt for greater benefits profitability and sustainability in larger areas and on more crops. Capacity building through training, creating awareness among producers, traders and consumers is necessary. Determination and desires to be the future agricultural policy to make India organic and the motto, grow organic, buy organic and eat organic.

REFERENCES

Angelina -Briones. 1999. Organic farming: Root of sustainable in Asia. In: *Food security in harmony with nature*. Proceedings of the 3rd IFOAM-Asia Scientific Conference and General Assembly. UAS, Bangalore, p.73-76.

Anonymous. 2004. Annual Progress Report of Network Project on Organic Farming.

Ghosh Saien. 1999. Principles of Organic Farming in the Tropics. *The organic farming readers*. Other India Press. p.53-68.

Kannaiyan, J., Tamil Selvan C; Muthaiyan M.C. and David B.V. 2005. Bioproducts in Organic Agriculture. Proceedings of the seminar on Organic agriculture in Peninsular India, OASIS, Coimbatore, p.109-113.

Menon, Manoj Kumar. 2007. The market potential for organic foods in India Organic Farming *News Letter* 3:22-26.

Narayanan, A. 2005. Understanding organic Agriculture. Proceedings of the seminar on Organic Agriculture in peninsular India,OASIS, Coimbatore, p11-14.

Palaniappan, S.P.and Annadurai, K. 1999. Organic farming; Theory and Practice. Scientific Publishers (India), Jodhpur, p18.

Rajendran, T.R., Venugoplam, M.V. and Tarhalkar, P.P. 1999. Organic Cotton Farming Technical Bulleting No.1/2000, CICR, Nagpur, 37.

Reganold, J.P., Palmer, A.S., Lockhart, J.C. and Macgregor, A.N. 1993. *Science*, 260, 344-349.

Selvaraj, N., Anitha, P., Ramaraj, P. and Mohan, L. 2005. Modern Organic Farming for Horticultural Crops (Tamil). Horticultural Res. Station, TNAU, Ooty.

Tiwari, R.C., Sharma, G.S. and Khandelwal, S.K. 2009. Organic Farming for Sustainable Agriculture, Himanshu Publication New Delhi 24-35.

❑❑❑

Chapter **16**

Organic Agriculture and GM Technology

Binay Kumar Singh and Lijo Thomas

Directorate of Rapeseed-Mustard Research, Bharatpur - 321 303, Rajasthan

INTRODUCTION

International Federation of Organic Agriculture Movements (IFOAM), an international umbrella organization for the organic movement, defines organic agriculture as "a production system that sustains the health of soils, ecosystems and people. It relies on ecological processes, biodiversity and cycles adapted to local conditions, rather than the use of inputs with adverse effects. Organic agriculture combines tradition, innovation and science to benefit the shared environment and promote fair relationships and a good quality of life for all involved". Clearly, this definition, which has been arrived at through a broad-based democratic process, does not restrain from using science, but is opposed to the application of certain modern bio-technologies such as Genetically Modified (GM) technology in agriculture. IFOAM believes that the application of GM technology in agriculture causes, or may cause severe environmental and social impacts including:

- Irreversible release of Genetically Modified Organisms (GMOs) that have never before existed in nature
- Pollution of the gene pool of cultivated crops, micro-organisms and animals
- Pollution of non-farm organisms
- Denial of free choice, both to farmers and consumers
- Violation of farmers' fundamental rights and endangerment of their economic independence
- Unacceptable threats to human health etc.

In their present form, the objections raised by IFOAM, however, is based more on a different perception of risks in using GM technology in agriculture than upon the presence or absence of scientific proof. As a result, stakeholders and experts characteristically advocate conflicting opinions. The proponents of organic agriculture encourage farmers to stop production and use of GM plants and implement organic farming. However, because of certain concerns over organic agriculture, mainly its lower productivity and therefore lower income, farmers, especially in developing countries feel insecure in accepting this environmentally-friendly approach. In developed countries, both the approaches can be seen to exist and operate independently. However, in the developing countries, conditions seem to be less favourable for the practice of organic farming. Under such circumstances, need based use of GM crops may be a complementary approach to organic agriculture. Though, there may be strict 'no' for this by the proponents of organic agriculture, the unbiased masses would certainly like to investigate the advantages offered by the GM crops as well as the organic crops, issues and concerns related to them, and the potential benefits and ways of combining the best aspects of the two technologies to maximise yields while limiting damage to ecosystems.

GM TECHNOLOGY

Benefits of GM Technology

The application of GM technology for agricultural use has been a subject of controversy since its inception. However, due to its great potential to ensure future food, feed and fibre security, especially for small-scale agriculture in developing countries, it is increasingly gaining faith and recognition of common man as well as policy makers around the world. The potential promise of GM technology to enhance the productivity of the crops on the current 1.5 billion hectares of arable land will have far reaching impact on food security as well as conservation of the environment. The production of renewal resource-based bio-fuels with the help of biotechnology will reduce dependence on fossil fuels. Thus, it would also be helpful in making the environment cleaner and safer with lower levels of greenhouse gases, necessary to mitigate global warming. GM supporters also argue that it can provide healthier, cheaper, and more stable crops which will benefit both the consumers and farmers. It is also envisaged that GM technology can play a major role in reducing the crop growing duration, while requiring less external inputs for managing soil fertility as well as biotic and abiotic stresses. Supporters of GM crops also acknowledge the potential of plant biotechnology to improve plant breeding strategies and designing new foods with specific health protective properties.

Concerns about GM Technology

The major concerns associated with the introduction and proliferation of GM technology are (i) Possible negative impacts on human and livestock health (ii) Impacts on environment, and (iii) Social and ethical issues associated with the technology. The potential environmental impacts include loss of flora and fauna biodiversity by extensive cultivation of GM crops and the possibility of the development of (a) super-weeds (b) super-pests and, (c) new viral strains, by the unintended gene flow through pollen to the sexually compatible plants. Likewise, the potential risks to the human and animal health include (a) allergenicity or toxicity and, (b) acquired antibiotic resistance by the human and animal pathogens by the horizontal transfer of antibiotic resistance gene. The ethical and social issues are varied and region or country specific. These are (a) selective adoption of this technology leading to further economic disparities, (b) monopolization of trade through operation of rigid intellectual property rights and restrictive trade practices, (c) skewness of new advances to interests of rich countries, (d) bio-piracy, or foreign exploitation of natural resources, (e) violation of organisms' intrinsic values, (f) tampering with nature by mixing genes among species, (g) objections to using animal genes in plants and vice versa and, (h) labelling issues related to GM products.

Many of these concerns though serious in nature are not based on scientific evidence and in some cases are merely exposition of possibilities. But at the same time, it is also true that, the absence of evidence cannot be the evidence for absence. Many of the objections raised against GM technology may seem irrational but it should be accepted that the present state of scientific knowledge is insufficient to prove or disprove them. A brief review of the course of evolution of the GM technology is appropriate before we undertake further analysis of the pros and cons of the technology.

GM CROPS: MYTH OR REALITY

In the early 1990s, opponents of crop biotechnology were cynical that GM crops, could ever deliver products able to make an impact at the farm level. Their reservations, however, received a big blow in the year 1996 when four transgenic crops comprising three insect-resistant crops and a herbicide tolerant soybean, developed by Monsanto Company, received regulatory approvals and these were commercially grown and harvested for the first time in the USA. Consequent to the realization of the economic, environmental and welfare benefits gained by the GM crops, small and resource-poor farmers around the world, particularly in the developing countries of Asia, Latin America and Africa, started adopting GM technology to their benefit. The data on GM cultivation and it spread in these countries acts as a pointer to this fact. Despite the on-going debate on GM crops, particularly in countries of the European Union, millions of large and small farmers in both industrial and developing countries continue to increase their plantings of GM crops because of the significant multiple benefits they offer. This unprecedented adoption rate reflects a strong vote of confidence and grower satisfaction in GM crops.

In 2008, the thirteenth year of commercialization of GM crops, of the 25 countries growing 125 million hectares of GM crops, 15 were developing countries, compared with only 10 industrial countries; and the trend in favour of developing countries is expected to continue in the future as more and more developing countries from all three continents in the South embrace GM crops. In 2008 alone, 3 new developing countries, Burkina Faso and Egypt from Africa, facing extreme technological, infrastructure and climatic challenges, and Bolivia from Latin America, joined the brigade of GM crop growers. Notably, of the 13.3 million farmers who benefited from GM crops in 2008, 90%, (approximately 12.3 million) were small and resource-poor farmers in developing countries and remaining were large farmers from developed countries such as the USA and Canada as well as developing countries including Argentina and Brazil.

CONTRIBUTION OF GM CROPS TO SUSTAINABILITY

Globally, GM crops have been quite successful in contributing to the sustainability by addressing the food requirements of people as well as concerns related to the environment.

FOOD SECURITY, ALLEVIATION OF POVERTY AND HUNGER

Nearly 50% of the poorest people of the world are small and resource-poor farmers, and another 20% are the rural landless people completely dependent on agriculture for their livelihoods. Majority of these farmers, mostly from developing countries, have not yet been able to realise the benefits of Green Revolution because its agricultural practices require significant capital investment, irrigation, modern farm machinery, fuels, agro-chemicals etc which are beyond their financial reach. Contrary to this, cultivation of GM crops is knowledge intensive as its technologies are incorporated into the seeds of the crop and are thus less dependent on external inputs. This unique quality of GM crops enables even the small and resource poor farmers to harness their benefits in terms of increased crop productivity under less cost. During the year 1996, the first year of commercialization of GM crops, to 2007, economic gains of 44.1 billion US dollar was made globally, 44% of which were due to substantial yield gains and 56% due to reduction in production costs. During 1996-2007, the cumulative farm income gain derived by the farmers of the developing countries was nearly 22.1 billion US dollar (50.1% of the total). In 2007 alone, the direct global farm income benefit from GM crops was approximately 10.1 billion US dollar about 58% of which went to developing countries. During 1996-2007, the global production gains were to the tune of 141 million tons which would have required 43 million additional hectares had GM crops not been deployed. These data indicate that GM technology has started bearing fruits and has enormous potential for the future, especially, when the GM version of staple crops like rice and wheat, as well as pro-poor drought tolerant food crops like cassava come to field.

BIODIVERSITY CONSERVATION

As mentioned earlier, GM technology has a great potential to increase crop productivity thereby, satisfying the needs of the people from relatively less land. It is estimated that, to feed the ever increasing population, approximately 13 million hectares of biodiversity rich forests are lost in developing countries every year. Though, under current rate of population growth, it would be difficult to check it completely, but by using this technology some restriction can be put to deforestation. Data indicates that during 1996-2007, by cultivating GM crops 141 million tons of additional food grains were produced which would have required an additional area of 43 million hectares of crop land had GM crops not been grown.

IMPROVEMENT OF ENVIRONMENTAL HEALTH

In the last one decade, global agriculture has witnessed a significant reduction in the consumption of pesticides, and savings in emissions of green house gases particularly CO_2 by the cultivation of GM crops. The savings in CO_2 emissions could have been possible firstly, due to reduced use of fossil based fuels, associated with fewer insecticides and herbicide sprays and secondly by additional soil carbon sequestration due to conservation tillage, facilitated by herbicide tolerant GM crops. Besides, reducing emissions of CO_2, conservation tillage has also helped in conserving soil as well as moisture, which has a major impact on sustainable cropping systems world wide, particularly in African countries where soil conditions are very poor and drought is more prevalent and severe than any other parts of the world.

It was estimated that during 1996-1997 an accumulative reduction of 359,000 metric tons of pesticides active ingredient (a.i), a saving in 9% in pesticides, was made. The corresponding data for 2007 alone was a reduction of 77,000 metric tons a.i (equivalent to a saving of 18% in pesticides. The savings in CO_2 emissions in the year 2007, through reduced use of fossil fuels as well as soil carbon sequestration was equivalent to a saving of 14.2 billion kg of CO_2.

ORGANIC FARMING

The concept of organic agriculture is not new, but the rising concerns among the public over pollution, food safety and human and animal health in the last few decades have invigorated interest in it. Today, consumers, particularly in developed countries, are willing to pay price premiums of 10-40 per cent for organic produce. As a result, organic agriculture has expanded rapidly in western countries. In 2001, some 15.8 million hectare area was under certified organic agriculture globally. Almost half of this was in Oceania, just under a quarter in Europe and a fifth in Latin America. About two-thirds of the area is organic grassland. As a percentage of total agricultural land, organic area is still modest an average of 2 per cent in Europe. However, many European countries have ambitious targets for expansion, with the result that Western Europe may have around a quarter of its total agricultural land under organic management by 2030. In many industrial countries, sales are growing at 15 to 30 per cent a year. Demand is expected to continue to increase, perhaps even faster than the 20 per cent or 50 per cent achieved in recent years.

The global market for organic food is worth $ 37 billion and is still growing, topically the United States, considered a world leader in transgenic technology, is the largest importer of organic foods, followed by the European Union. The shift in consumer preference for organic products is spreading worldwide. According to the World Health Organization (WHO), the $ 14 billion global market for organic herbal plants and medicines is growing at 15-25 per cent a year and is likely to be worth a mammoth $ 5 trillion in 2050. However, India's export of organic food is worth hardly Rs. 90 crores. Organically produced agricultural products from India like tea, coffee, spices, fruits, vegetables, cotton, rice, oilseeds, pulses and sugarcane has already found markets and high degree of acceptability in several countries. Though organic agriculture accounts only for a negligible portion of the country's total farm produce, there is a great demand for it in the international market.

ADVANTAGES OF ORGANIC FARMING

Organic farming offers several benefits to the society; the most obvious advantage that might mostly be considered for small-scale farmers is their traditional knowledge of the natural environment that can be well-matched with organic farming. Concurrently, organic farming avoids chemical inputs ie. fertilizers, insecticides, herbicides, which not only put significant financial burden on the farmers, but also acts as potential source of pollution in the environment. Organic farming also helps to improve soil fertility, by promoting green coverage, and promote biodiversity. Thus, as a whole organic farming may be presumed as the most 'environment friendly' farming system.

CONCERNS ABOUT ORGANIC FARMING

In spite of all good things and plentiful advantages that organic farming offers to the society, its practice by the developing countries may still encounter some difficulties. To date, however, it is difficult to say whether or not organic farming is a low cost technology because production cost of crops is often higher by the adoption of organic farming, more especially if major soil protection or restoration activities are needed. Limited availability of institutional scientific research on organic farming methods; lack of access to organic planting materials, plant protection inputs and animal breeds; poor knowledge and ability to react to unpredicted external factors, such as drought, the sudden arrival of new diseases or pests; high certification cost; little bargaining power with buyers; poorly developed markets for organic produce and the bias of most legal structures in favour of conventional agriculture and several other factors associated with small-scale farmers are also detrimental to the adoption of organic farming. Transition to organic farming from almost centuries old conventional agriculture is also not easy as it unavoidably increase the prices of the products and therefore, low income families and people, in developing countries, despite their desire, cannot afford it. Finally, expansion of organic farming to the level to feed the entire world is practically not possible as extra land to compromise its low productivity per unit area cannot be made available.

CO-EXISTENCE OF GM TECHNOLOGY AND ORGANIC FARMING

From the above discussions, it is clear that the benefits of GM technology are enormous and will be useful to the growing world population. However, the uncertainty in the environmental effects is a major problem concerning the GM technology, especially in relation to the pest resistant GM crops. There are also some concerns related to the loss of biodiversity and other environmental risks effecting the ecosystems, agriculture and health. Though, with the present state of knowledge it is very difficult to comprehend and predict these effects, careful and judicious application of GM technology may surely reduce this risk. Overall, many researchers believe that the benefits of GM technology are more than their negative consequences. Therefore, before any interventions in the complicated third world's farming systems, risks of producing and consuming new GM foods should be evaluated against potential benefits, and when benefits are much bigger than the risks, such foods can be introduced and adopted.

On the other hand while many farmers in the developing countries are still practicing agriculture without the use of any chemical inputs fertilizer and pesticides, they are not organic farmers as they do not benefit from the organic farming techniques. Subsistence farmers in these countries have different circumstances than farmers in the developed countries. Their main concern is the survival of the family rather than the profit from national and international market.

Experts believe that while there is a great potential for organic farming in South-East Asia, East Europe and the countries around the Baltic Sea, little exits in the countries of Africa, even if organic principles and techniques may be useful. The demand market for organic crops is very small in Africa, and if the organic farming became certified with a possibility to sell to Europe or North America, it would only be a few products in a small scale. Also, the soil conditions in most of the African countries are very poor and climatic conditions are such that soil organic matter is degraded very fast. Thus, most aid for these countries should be focused on increasing production and international trade rather than organic farming methods.

Thus, it is evident that circumstances are different in different countries and therefore, organic agriculture alone may not be able to serve the purpose of these countries. Therefore, appropriate measure would be to combine the best aspects of the two technologies to harness maximum benefits. While, GM technology would help in designing crops suited to their agro-climatic conditions, implementation of organic agriculture will reduce the environmental hazards caused due to conventional agriculture practices. However, it would certainly require relaxation in the principles of organic farming to accommodate some of the very promising products of GM technology.

CONCLUSION

GM technology has enormous benefits to offer to the society, particularly to the countries of the south where the agro-climatic conditions are tough and unfavourable for farming. The growing number of the countries joining the brigade of GM crop users and enormous gain in the public confidence ensures the existence of GM technology in the future. Government and policy makers in the developing countries are therefore, facing a crucial question on how GM technology should be viewed in relation to organic farming. Though, organic farming has several environmental and health benefits, it is finding hard to sustain in the developing countries because of generally poor soil conditions, lack of organic matter, labour intensive nature of the technology, increased cost, invisibility of results in short-term and virtually no access to international markets. Under such circumstances a two pronged approach combining GM technology with organic farming may well be the best way to resolve the need for increased global food production, while minimizing environmental impact. Its time to forget about the ideological warfare, since the biotech crop industry has gained in expertise and learned how to deal with environmental problems. So – in essence – we should learn to work together and plan for organo-transgenic crops to exploit maximum benefits in a sustainable manner.

REFERENCES

Brookes, G. and Barfoot, P. 2009. GM crops: global socio-economic and environmental impacts 1996-2007. PG Economics Ltd, UK.

Cacek, T. and Langner, L.L. 1986. The economic implications of organic farming. Available on: http://www.eap.mcgill.ca/MagRack/AJAA/AJAA_2.htm.

Clive, James 2008. Global status of commercialized biotech/GM crops. ISAAA, Brief No. 39.

Closter, A.M., Embree, E., Petersen, M. and Pedersen, S.H. 2004. Organic farming and genetically modified crops in relation to food security. Henning Høgh Jensen Publication.

Foley, C.L. 2006. Making the organic connection. Available on: http://www.organicconnection.net/.

Foodaware. 2002. Organic farming and foods. Available on: http://www.net-consumers.org/food/03_02rev.htm.

Greenthumb. 2002. Twenty-one year study documents the many benefits of organic agriculture. Available on: http://www.npsas.org/BenefitsOrg.html.

http://ifoam.org/growing_organic/definitions/doa/index.html

Jensen, H.H. 2004. Tropisk, okologi, Globale, Gener. Available on: http://www.globalegener.dk/ulandeoggentek/tropisk.asp.

Lampkin, N. and Midmore, P. 1999. Organic farming and the European Union. Memorandum of evidence to the House of Lords select committee on the European communities sub-committee D (Agriculture, Fisheries and Food). Available on: http:// www.aber.ac.uk/~wirwww/organic.

Organic Farming Research Foundation 2006. Frequently asked questions about organic farming. Available on: http://www.ofrf.org/general/about_organic/.

Parrott, N. 2004. Sound depths: the seen and unseen dimensions of organic farming in the south — and their implications for organic research. Paper to DARCOF Workshop on "*Organic Farming in a Global Perspective: Globalization, Sustainable Development and Ecological Justice*".

Pretty, J.N., Morrison, J.I.L., Hine, R.E. 2002. Reducing food poverty by increasing agricultural sustainability in developing countries. *Agric Ecosyst Environ*, 95:217–34.

Uzogara, S.G. 2000. The impact of genetic modification of human foods in the 21st century: a review. *Biotechnol Adv*, 18:179–206.

❑❑❑

Chapter **17**

Certified Organic Farming : Evolution and Regulations

S.S. RATHORE* AND KAPILA SHEKHAWAT**

**Senior Scientist, Agronomy, Directorate of Rapeseeed-Mustard Research, Bharatpur, Rajasthan*
***Scientist, Agronomy, Directorate of Rapeseeed-Mustard Research, Bharatpur, Rajasthan.*

In ancient and medieval time and more precisely prior to the 19th century, most food in the world was organically produced. The invention of horse-drawn hoe and a seed drill by Jethro Tull began the agricultural revolution in England during early 19th century. The first category of chemicals being discovered was the single superphosphate, urea and DDT and was followed by the discovery of BHC in France. Nitrophenols were the first group of selective herbicides developed in 1933 and were followed by the development of 2, 4-D and MCPA in 1940s. India's own achievements in agricultural production after the Green Revolution in the mid-1960s, has been exemplary and mainly due to increased use of the components of modern agriculture, namely fertilizer, pesticides and farm machinery. But the rising awareness towards organic agriculture in present time is due to the fact of biological magnification of pesticides and fertilizer residues in food chain and degradation of soil natural fertility status. Injudicious use of inorganic nutrient sources and complete negligence for inclusion of organic sources as a part of fertility management has reduced the soil resilience to an ever time low level. These ill-effects of modern agriculture forced people, especially in countries with high-income economies (European countries, USA, Canada, Australia, etc.) to demand food grown without fertilizers and pesticides. This paved the way for organic farming. Continued use of organic manure on a farm improves its organic matter content, which supports the soil micro, meso and macro fauna and makes the soil a living body (Ramakrishnan,*et.al.*, 2003). A large number of terms are used as an alternative to organic farming like biological, ecological, bio-dynamic, organic–biological and natural agriculture.

DEVELOPMENTS IN CERTIFIED ORGANIC FARMING

Although the term organic farming is getting popularity in recent times but it was initiated in 10000 years back when ancient farmers started cultivation depending up on natural sources only. In fact, organic agriculture has its roots in traditional agriculture practices that evolved over the millions. Major milestones in the area of organic farming are presented in the table 1.

Table 1 : Historical development of organic farming in the world

Ancient period	
Oldest practice	10000 years old, dating back to Neolithic age practiced by ancient civilizations like Mesopotamia, Hwang Ho basin etc.
Ramayana	All dead things- rotting corpse or stinking garbage returned to earth are transformed into wholesome things that nourish life- as interpreted by C. Rajgopalachari.
Mahabharata (5500 BC)	Mention of kamdhenu, the celestial cow and its role of human life and soil fertility.
Kautilya Arthashatra	Mention several manures like oil cake, excreta of animals
Brihad-Sanhita (By Virahmihir)	Described how to choose manures for different crops and the methods of manuring
Holy Quran (590 AD)	At least one-third of what we take out from soil must be returned to it implying recycling or post-harvest return.
Key milestones in organic farming current period	
Sir Albert Howard (1900-1947)	Father of modern organic agriculture developed organic composting process (*Mychorrizal* fungi) at Pusa, Samastipur, India and published document, "An agriculture Testament"
Rudolf Steiner (1922)	A Germen Philosopher developed biodynamic farm in Germany
J.I Rodel (1950) USA	Popularised the term sustainable agriculture and methods of organic growing
IFOAM	Establishment of International Federation of Organic Agriculture Movement in 1972
One straw regulation	Realease of the book by Masanabu Fukoka, an eminent microbiologist in Japan
EU regulation	EU regulations on Organic Foods, 1991
CODEX	CODEX guidelines on organic standards, 1999

Source: Bhattacharyya and Chakraborty, 2005

DEVELOPMENT OF ORGANIC PRODUCT CERTIFICATION IN INDIA

India ranks 33rd in terms of total land under organic cultivation and 88th position for agriculture land under organic crops to total cultivated area. The cultivated land under certification is around 2.8 million ha (2007-08). This includes 1 million ha under cultivation and the rest is under forest area (APEDA).The four pillars of organic farming are: organic standards, certification/ regulatory mechanism, technology packages and marketing network. The most important aspect of organic farming is certification which consists of standards, inspection and certification. After this certification, the product can be sold with an "organic" product label. The certification is valid only from accredited certifying agencies.

In India Organic farming started as below:

- Indiscriminate and excessive use of chemicals during post Green Revolution period has put forth a question mark on sustainability of agriculture in the long run. Organic farming appears to be one of the options for suitability. Starting of organic agriculture in India from 1900 by Sir Albert Howard, a British agronomist in North India, development of Indore method of aerobic compost (Howard, 1943), Banglore method of anaerobic compost (Acharya, 1934) initiated organic agriculture in India.
- The planning Commission constituted a steering group on agriculture who identified organic farming as national challenge and suggested it should be taken in the form of a project as a major thrust area for 10th Plan.
- The National Agricultural policy (2000) recommended promotion of traditional knowledge of agriculture relating to organic farming and its scientific up gradation.
- The department of Agriculture and Cooperation (DAC), Ministry of agriculture (2000), constituted a task force on organic farming under the chairmanship of Shri Kumwar ji Bhai Yadav and this task force recommended promotion of organic farming.
- The Ministry of Commerce launched the National Organic Programme in agriculture and Agricultural and Processed Food Products Export Development Authority (APEDA) has implemented the National Programme on Organic Production (NPOP). Under NPOP, documents like National standards, accreditation criteria for accrediting inspection and certification agencies, Accreditation procedures have been prepared and approved by National Steering Committee (table-2).

Table 1 : List of accrediting, inspection and certification agencies of India under NPOP

S. No.	Name of the agency	Address	Contact
1.	BVQI (India) Pvt. Ltd.	Marwah Centre, 6th Floor, Opp. Ansa Industrial Estate Krishanlal Marwah Marg Off Saki-Vihar Road ndheri (East)Mumbai	Contact Person: Mr. R. K. Sharma Phone No.: 022-56956300, 56956311 Fax No. 022-56956302 / 10 Email: scsinfo@in.bureauveritas.com
2.	Ecocert SA (India Branch	Sector-3, S-6/3 & 4, Gut No. 102 Hindustan	Contact Person: Dr. Alexander Daniel Phone No.: 0240-2377120, 2376949

contd. ...

... contd,

S. No.	Name of the agency	Address	Contact
	Office) (Maharashtra)	Awas Ltd. Walmi-Waluj Road Nakshatrawadi Aurangabad – 431 002	Fax No.: 0240-2376866 Email: ecocert@sancharnet.in
3.	IMO Control Private Limited	No. 1314, Double Road Indiranagar 2nd Stage Bangalore-560 038. (Karnataka)	Contact Person: Mr. Umesh Chandrasekhar Phone No.: 080-25285883, 25201546 Fax:080-25272185 Email: imoind@vsnl.com
4.	Indian Organic Certification Agency (INDOCERT))	Thottumugham P.O. Aluva-683 105 Cochin, (Kerala	Contact Person: Mr. Mathew Sebastian Telefax:0484-2630908 Email: Mathew.Sebastian@ indocert.org
5.	International Resources for Fairer Trade Sona Udyog (Industrial Estate)	Unit No. 7, Parsi Pandhayat Road Andheri (E) Mumbai – 400 072 (Maharashtra)	Contact Person: Mr. Arun Raste Phone No.: Tel: 022-28352811, 28235246 ext.22Fax–022-823-5245 Email: irft@vsnl.com
6.	Natural Organic Certification Association	Chhatrapati House Ground Floor Near P. N. Gadgil Showroom Pune-411 038 (Maharashtra)	Contact Person: Mr. Sanjay Deshmukh Phone No.: 020-25457869, 56218063 Fax: 020-2539-0096 Email: contact@nocaindia.com
7.	OneCert Asia Agri Certification Private Limited	Agrasen Farm, Vatika Road, Off Tonk, Jaipur-303 905, (Rajasthan)	Contact Person: Mr. Sandeep Bhargava Phone No. : - 0141-2720202 to 0141-2770342 Telefax No: - 0141-2720202 Email: info@onecertasia.in
8.	SGS India Pvt. Ltd. -	250 Udyog Vihar Phase – IV Gurgaon – 122 015 (Haryana)	Contact Person: Mr. Sudarshan Sharma Phone No.: 95124-2399990 to 98 Fax No.: 95124-2399764 Email: sudarshan_sharma@sgs.com

contd. ...

... contd,

S. No.	Name of the agency	Address	Contact
9.	Skal International (India) A Division of CU Inspections India Pvt. Ltd.	No. 191, 1st Main Road Mahalaxmi Layout Bangalore – 560 086 (Karnataka)	Contact Person: Mr. Narayana Upadhyaya Phone No.: 080-23491928, 56966507 Fax no.: 080-23491935 Email: skalindia@eth.net
10.	Uttaranchal State Organic Certification Agency (USOCA)	12/II Vasant Vihar Dehradun-248 006 (Uttaranchal)	Contact Person: Dr. S. K. Malik Phone No.: 0135-2760861 Fax: 0135-2760734 Email: uss_opca@rediffmail.com

Source: APEDA

- National Steering Committee comprising Ministry of Commerce, Ministry of Agriculture, APEDA, Spice Board, Coffee Board, Tea Board and various other government and private organizations associated with organic movement is monitoring the overall organic activities under the NPOP.

India is in a very nascent stage when it comes to the organic exports, but the area and products are increasing under organics (table-3).

Table 3 : Scenario of organic farming trade in India (2004-2005)

Area under certification	2.54 mha
Total number of products certified	35
Total certified product	115,238 metric tons
Estimated organic value	710 million INR
Organic exporters:	• Sunstar overseas Ltd. Haryana • Navdanya Trust, New Delhi • FAB India Ltd., New Delhi • Kashmir Apiaries Export • Jubilant Organosys Ltd., Noida • IITC Organic India Ltd., Lucknow
No. Of processing units	158
Accredited I & C agencies	11
Total Quantity exported	6,472 MT
States involved in organic exports (tonnes)	
Kerela	1232
West Bengal	937

contd. ...

... contd.

Karnataka	476
Tamil Nadu	471
Punjab	541
Himachal Pradesh	521
Maharashtra	375
Total	6472

Source : APEDA, 2004-05

Codex Alimentarius Commission, a joint body of FAO/WHO defines "organic agriculture as holistic food production management systems, which promotes and enhances agro-ecosystem health, including biodiversity, biological cycles and soil biological activity.

Organic production concept needs a well set of inputs, management techniques which are established and approved by many countries of the world in the form of standards. The formation of the International Federation of Organic Agriculture Movements (IFOAM) in 1972, gave an international framework for the discussion and codification of internationally recognized principles of organic farming.

Organic farming requires certification by accredited agencies in India, like IFOAM, Tea Board, Coffee Board, Spices Board, Coconut Development Board and Directorate of Cashew and Cocoa. The steps involved in certification are: registration of the producers and processing industries, provision of basic information on the crop and farm, inspection and verification of the field and processing unit, inspection of production methods and practices by the inspector of the certifying agency. A transition period of 2–3 years is required during which no inorganic fertilizer and agrochemicals are used on the farm before which the produce can be marketed.

Department of Commerce, Ministry of Commerce and Industry has formulated sound programme for certified organic farming in India under the aegis of "National Programme for Organic Production (NPOP)," which was first published during May, 2001 and since than six editions have added for improvement. Under this programme , it is better to understand some of the terms for sound knowledge of organic farming certification.

Accreditation

Accreditation means Registration by the National Accreditation Body for certifying organic farms, products and processes as per the National Standards for Organic Products and as per the guidelines of the National Accreditation Policy and Programme for organic products. In India, Agricultural and Processed Food Export Development Authority (APEDA) is one of the accreditation agencies for organic certification.

Certificate

Certificate would mean a document issued by an accredited agency declaring that the operator is carrying out the activities or the stated products have been produced in accordance with the specified requirements to the National Standards for Organic Products. A certificate shall be the procedure by which a written assurance is given by the Certification Agency that a clearly identified production or processing system has been methodically assessed and conforms to the specified requirements.

The National Programme for Organic Production proposes to provide an institutional mechanism for the implementation of National Standards for Organic Production, through a National Accreditation Policy and Programme. The aims of the National Programme for organic production include the following:

(a) To provide the means of evaluation of certification programmes for organic agriculture and products as per the approved criteria.

(b) To accredit certification programmes.

(c) To facilitate certification of organic products in conformity to the National Standards for Organic Products.

(d) To encourage the development of organic farming and organic processing.

The National Steering Committee of NPOP would also function as the National Accreditation Body. The members of the National Accreditation Body shall comprise of representatives from Ministry of Agriculture, Ministry of Commerce and Industry, APEDA, Coffee Board, Spices Board and Tea Board.

HAZARD ANALYSIS CRITICAL CONTROL POINT (HACCP)

The Hazard Analysis Critical Control Point (HACCP) system is based on systematic approach to produce safe food. Food safety management systems based on HACCP are internationally recognized as the most effective way to ensure food safety and minimize the risks of food poisoning.

NATIONAL STANDARDS FOR ORGANIC PRODUCTION

Ministry of Commerce Under the "National Programme for Organic Production" has prescribed National Standards for Organic Production. These standards are grouped under following six categories:

1. Conversion
2. Crop production
3. Animal husbandry
4. Food processing and handling
5. Labeling
6. Storage and transport

CONVERSION REQUIREMENTS

The time between the start of organic management and cultivation of crops or animal husbandry is known as the conversion period. The whole farm including the livestock should be converted to the standards over a period of time. All standard requirements should be met during conversion period. If the whole farm is not converted then the two must be separated and inspected. Regular inspections during the conversion period should be carried out.

Simultaneous productions of conventional or in conversion and or organic which can not be distinguished clearly are not allowed. To ensure clear separation the certification programme shall inspect the whole production system. Full conversion period is not required where organic farming practices are already in use. But this has to be verified by the inspection agency. As per organic certification directory, 2003 there are 364 certifying agencies in the world.

Maintenance of organic management

Organic certification is based on continuance. The certification programme should certify the production, which is likely to be maintained on a long term basis. The converted land and animals shall not get switched back and forth between organic and conventional management.

Landscape

The certification programme shall set standards for a minimum percentage of the farm area to facilitate biodiversity and nature conservation. Organic farming should contribute beneficially to the ecosystem. Areas which should be managed properly and linked to facilitate biodiversity are:

- Extensive grassland
- All areas which are not under rotation and are not heavily manured
- Pastures, meadows, orchards, hedges, hedgerows etc
- Ecologically rich fallow land or arable land
- Ecologically diversified field margins
- Waterways, pools, springs, ditches, wetland, swamps

CROP PRODUCTION

Choice of crops and varieties—All seeds and planting materials should be certified organic, well adapted to local climatic conditions and resistant to pests and diseases. If certified organic seed or planting material is not available then chemically untreated conventional material can be used. Uses of genetically engineered seeds, pollen, transgenic plants are not allowed.

Duration of conversion period—The minimum conversion period for plant products, produced annually is 12 months prior to the start of the production cycle. For perennial plants (excluding pastures and meadows) the conversion period is 18 months from the date of starting organic management. Depending upon the past use of the land and ecological situations, the certification agency can extend or reduce the minimum conversion period.

Diversity in crop production—Diversity in crop production is achieved by a combination of (a) versatile crop rotation with legumes and (b) by appropriate coverage of the soil with diverse plant species during the year of production that, taken into account pressure from insects, weeds, diseases and other pests, while maintaining or increasing soil health and fertility.

Fertilization policy—The organic philosophy of crop nutrition begins with proper care and nourishment of the organism responsible for the soil digestive process. It avoids excessive tillage and use of toxic chemicals that are harmful for soil organisms as well as addition of organic matter and natural rock minerals. Biodegradable material of plant or animal origin produced on organic farms should form the basis of the fertilization policy. Fertilization management should minimize nutrient losses, avoid accumulation of heavy metals and maintain the soil pH. Emphasis should be given to generate and use own on-farm organic fertilizers. Brought in fertilizers of biological origin should be supplementary and not a replacement. Over manuring should be avoided. Manures containing human excreta should not be used on vegetation for human consumption.

In case of deficiency mineral fertilizers can be used as supplementary source and should be applied in their natural composition. Minerals containing high concentrations of heavy metals should be avoided. Biofertilizers can be used safely under all ecosystems and in all the crops. Table 4 represents the items which are permitted or restricted to be used in organic farming.

Table 4 : Products for use in fertilization and soil conditioning in organic farming

Items	Conditions for use
Material from plant and animal origin	
Matter produced on an organic farm unit	
• Farmyard and poultry manure, slurry, urine	Permitted
• Crop residues and green manure	Permitted
• Straw and other mulches	Permitted
• Composts and Vermicompost	Permitted
Matter produced outside the organic farm unit	
• Blood meal, meat meal, bone meal and feather meal without preservatives	Restricted
• Compost made from plant residues and animal excrement	Restricted
• Farmyard manure, slurry, urine	Restricted
• Fish and fish products without preservatives	Restricted
• Guano	Restricted
• Human excrement	Restricted
• Wood, bark, sawdust, wood shavings, wood ash, wood charcoal	Restricted
• Straw, animal charcoal, compost and spent mushroom and vermiculate substances	Restricted
• Compost from organic household	Restricted
• Compost from plant residues	Restricted
• Sea weed and sea weed products	Restricted
By products from the industries	
• By-products from the food and textile industries of biodegradable material of microbial, plant or animal origin without any synthetic additives	Restricted
• By products from oil palm, coconut and cocoa (including fruit bunch, palm oil mill effluent, cocoa peat and empty cocoa pods.	Restricted

contd. ...

... contd.

Items	Conditions for use
• By-products of industries processing ingredients from organic agriculture	Restricted
• Extracts from mushroom, *Chlorella*, Fermented product from *Aspergillus*, natural acids (vinegar)	Restricted
Mineral Origin	
• Basic slag	Restricted
• Calcareous and magnesium rock	Restricted
• Lime, limestone, gypsum	Restricted
• Calcified sea weed	Restricted
• Calcium chloride	Restricted
• Mineral potassium with low chlorine content (e.g. sulphate of potash, kainite, sylvinite, patenkali)	Restricted
• Natural phosphates (rock phosphate)	Restricted
• Trace elements	Restricted
• Sulphur	Restricted
• Clay (bentonite, perlite, zeolite)	Restricted
Microbiological origin	
• Bacterial preparations (biofertilizers)	Permitted
• Biodynamic preparations	Permitted
• Plant preparations and botanical extracts	Permitted
Annexure II Products for Plant pest and disease control	
Material from plant and animal origin	
• Plant based repellents (Neem preparations from *Azadirachta indica*)	Permitted
• Algal preparations (gelatin)	Permitted
• Casein	Permitted
• Extracts from mushroom, *chlorella*, fermented products from *Aspergillus*	Permitted
• Propolis	Permitted
• Beeswax, Natural acids (vinegar), plant oils, Quassia	Restricted
• Rotenone from Derris elliptica, *Lonchocarpus*, *Thephrosia* spp.	Restricted
• Tobacco tea (pure nicotine is prohibited)	Restricted

contd. ...

... contd.

Items	Conditions for use
• Preparation from *Ryania* species	Restricted
Mineral origin	
• Chlorides of lime/soda	Restricted
• Burgundy mixture	Restricted
• Clay (bentonite, perlite, vermiculite, zeolite)	Permitted
• Copper salts/ inorganic salts (Bordeaux mix, copper hydroxide, copper oxychloride)	Not allowed
• Quick lime	Restricted
Mineral origin	
• Diatomaceous earth	Permitted
• Light mineral oils	Restricted
• Permangnate of potash	Restricted
Insects origin	
• Release of parasites, predators of insect pests	Restricted
• Sterilized insects	Restricted
• Sterilized insect males	Not allowed
Microorganisms used for biological pest control	
• Viral, fungal and bacterial preparations (biopesticides)	Restricted
Others	
• Carbon dioxide and nitrogen gas	Permitted
• Soft soap, soda, sulphur dioxide	Permitted
• Homeopathic and ayurvedic preparations	Permitted
• Herbal and biodynamic preparations	Permitted
• Sea salt and salty water	Permitted
• Ethyl alcohol	Not allowed
Traps, barriers and repellants	
• Physical methods (e.g. chromatic traps, mechanical traps)	Permitted
• Mulches, nets	Permitted
• Pheromones – in traps and dispensers only	Permitted

Pest disease and weed management including growth regulators—Weeds, pests and diseases should be controlled by a number of preventive cultural techniques, such as suitable rotations, green manures, a balanced fertilization programme, early and pre-drilling seed bed preparations, mulching, mechanical control and the disturbances of pest development cycles.

Botanical pesticides prepared at farm from local plants, animals and microorganisms are allowed. Thermic weed control and physical methods for pests, disease and weed management are permitted. Use of synthetic chemicals such as fungicides, insecticides, herbicides, synthetic growth regulators and dyes are prohibited. Use of genetically engineered organisms or products is prohibited.

Contamination control—All attempts should be made to minimize contamination from outside and with in the farm.

Soil and Water conservation—Soil and water resources should be handled in a sustainable manner to avoid erosion, salination, excessive and improper use of water and the pollution of surface and ground water. Cleaning of land by burning (e.g. slash and burn and straw burning) should be restricted. Clearing of primary forest for agriculture (jhuming or shifting cultivation) is strictly prohibited.

Organic Certification Mark: organic Logo

In India, organic farming requires certification by accredited agencies like IFOAM, Tea Board, Coffee Board, Spices Board, Coconut Development Board and Directorate of Cashew and Cocoa. Agencies promoting organic farming in India are: National Steering Committee, Ministry of Commerce, Government of India and APEDA. APEDA is the coordinating agency for organic food processing and export under the brand name "India Organic". A trademark – "India Organic" will be granted on the basis of compliance with the National Standards for Organic Production (NSOP). Communicating the genuineness as well as the origin of the product, this trademark is owned by the Government of India. Only such exporters, manufacturers and processors whose products are duly certified by the accredited inspection and certification agencies, will be granted the licence to use of the logo which would be governed by a set of regulations.

ANIMAL HUSBANDRY

Maintenance\rearing of animals

The certification programme shall ensure that the management of animal environment takes into account the behavioral needs of the animal and provides for :

a. Sufficient free movement
b. Sufficient fresh air and day light
c. Protection against excessive sunlight, temperature, rain, wind etc.
d. Enough lying and resting area
e. Ample access to fresh water and feed and
f. Proper environment for their biological and ethological needs

Poultry and rabbits should not be kept in cages. Land less animal husbandry system shall not be allowed.

Length of conversion period—The whole farm including livestock should be converted to organic according to the standards. Animal products may be certified organic only after the farm has been under conversion for at least 12 months and the required standards have been achieved. Length of the conversion period can be extended at the discretion of the certification agency. In case of dairy and egg production the conversion period shall be 30 days at minimum.

Brought-in animals

All organic animals should be born and raised on the organic holding. When organic livestock is not available the certification programme shall allow brought-in conventional animals according to the specified age limits e.g. 2 days old chicken for meat production, 18 weeks old hen for egg production, 2 weeks old for any other poultry, piglets up to 6 weeks old after weaning and calves up to 4 weeks old which have received colostrums and are fed a diet consisting mainly of full milk.

Breeds and breeding

Breed should be chosen which are adapted to the local conditions. Breeding goals should not be in opposition to animal's natural behavior and be directed towards good health.

Artificial insemination is allowed. Embryo transfer techniques are not allowed. Hormonal heat treatment and induced births are not allowed unless applied for medical reasons. Use of genetically engineered species or breeds is not allowed.

Mutilations

Mutilations of animals in any form are not allowed. Certification programme may allow following exceptions – Castration, tail docking of lambs, dehorning, ringing and mule sing etc.

Animal nutrition

The livestock should be fed 100% organically grown feed of good quality. All feed should come from the farm itself or be procured from the region. The certification programme shall draw up standards for feed and feed ingredients.

Where it proves impossible to obtain certain feeds from organic farming sources, the certification programme shall allow a percentage of feed consumed by farm animals to be sourced from conventional farms subject to a maximum prescribed limit.

Synthetic growth promoters or stimulants, synthetic appetizers, preservatives, artificial colouring agents, urea, farm animal by products to ruminants, droppings, dung or other manure, feed subjected to solvent extraction (soy and rapeseed meal), pure amino acids and genetically engineered organisms or their products are strictly prohibited in the feeds.

Vitamins, trace elements and supplements shall be used from natural origin. Certification programme can define conditions for use of vitamins and minerals from synthesized or unnatural sources. For fodder preservation bacteria, fungi and enzymes, by products of food industry (such as molasses) and plant based products can be used.

FOOD PROCESSING AND HANDLING

General principles—Organic products shall be protected from co-mingling with non-organic products, and shall be adequately identified through the whole process. Certification programme shall regulate the means and measures to be allowed or recommended for decontamination, clearing or disinfection of all facilities where organic products are kept, handled, processed or stored. Besides storage at ambient temperature the following special conditions of storage are permitted: controlled atmosphere, cooling, freezing, drying and humidity regulation.

Pests and disease control—For pest management and control following measure shall be used in order of priority:

- Preventive methods such as disruption, and elimination of habitat and access to facilities.
- Mechanical, physical and biological methods
- Permitted pesticidal substances as per the standards and
- Other substances used in traps.
- Irradiation is prohibited.
- Direct or indirect contact between organic products and prohibited substances (such as pesticides) should not be there.

Ingredients, Additives and Processing Aids

100% of the ingredients of agricultural origin shall be certified organic. For the production of enzymes and other microbiological products, the medium shall be composed of organic ingredients.

In case where an ingredient of organic origin is not available, the certification programme may allow use of non-organic raw material subject to periodic re-evaluation. The same ingredient within one product shall not be derived both from organic and inorganic origin. Minerals, vitamins and similar isolated ingredients shall not be used. The use of additives and processing aids shall be restricted.

Preparations of microorganisms and enzymes commonly used in food processing can be used. But no genetically engineered microorganisms and their products shall be used.

Processing methods—Processing methods should be based on mechanized, physical and biological processes, so that the quality of organic ingredients is maintained through the process. Some of the approved processes are: mechanical and physical, biological, smoking, extraction, precipitation and filtration.

Extraction shall only takes place with water, ethanol, plant and animal oils, vinegar, carbon-dioxide, nitrogen or carboxylic acids and all these shall be of food grade quality.

Packaging—Material used for packaging shall be ecofriendly. Unnecessary packaging material should be avoided. Recycling and reusable systems should be used. Packaging material should be biodegradable. Material used for packaging shall not contaminate the food.

LABELING

When the full standard requirements are met, the product can be sold as "Organic". On proper certification by certification agency "India Organic" logo can also be used on the product.

STORAGE AND TRANSPORT

Products integrity should be maintained during storage and transportation of organic products. Organic products must be protected from co-mingling with non-organic products and must be protected all times from contact with the materials and substances not permitted for use in organic farming

CONCLUSION

The awareness for growing food through organic way is gaining importance worldwide. In India also, Government is promoting organic farming in organized manner. National programme for organic production is the flagship scheme of GOI, under which all rules and regulations for certified organic production have laid down. Some of the states in India especially in North east and Utttakhand certified organic production has already been started in big way. All arable land can not brought under organic production because issues like socio organic and food security reasons, but in areas of low productivity where use of chemicals in agriculture is minimal are the potential areas identified by the Task force set up by GOI. Conversions, Crop production, Animal husbandry, Food processing and handling Labeling, Storage are the six areas where standards have been set up under NPOP. Theses are not only the regulation but also wayforeward for start of certified organic production on sustainable basis for the growers, policy makers and researchers.

NPOP promotes organic farming in India and its biggest challenge is to arrange organic inputs to the farmers and provide low cost certification for higher profits. The areas should be targeted for promoting organic green food. Although, slowly but the major cultivable area should be transformed to chemical free organic zones for environmental safety and food swcurity.

REFERENCES

Bhattacharyya, P. and Chakraborty, G. 2005. Current status of organic farming in India and other countries. *Indian J. of fertilizers.* 1(9): 111-123.

Howard, A. 1943. An Agricultural Testament. Oxford University Press. London. 253p.

Ramakrishnan, P.S., Swift, M.J., Saxena, K.G., Rao, K.S., and Maikhuri, R.K. 2003. Soil Biodiversity, Ecological Processes and Landscape Management. Oxford & IBH, New Delhi.

❑❑❑

Chapter **18**

Socio-economic Impact of Organic Farming: Concept, Measurement and Issues

LIJO THOMAS* AND BINAY KUMAR SINGH**

** Scientist, Agricultural Economics, Directorate of Rapeseed-Mustard Research, Bharatpur, Rajasthan*

*** Scientist, Biotechnology, Directorate of Rapeseed-Mustard Research, Bharatpur, Rajasthan*

Indian farming has been bequeathed with a preponderance towards organic farming. Organic agriculture is getting significant attention from farmers, traders, consumers and policy makers alike in the recent years due to various factors. Organic farming has generated wide spread and passionate debate among social scientist also with regard to its socio-economic impact. Organic farming is a term with broad connotations and a precise definition acceptable to diverse section of academia may not be forthcoming. According to The International Federation of Organic Agriculture Movements (IFOAM) "Organic agriculture combines tradition, innovation and science to benefit the shared environment and promote fair relationships and a good quality of life for all involved." Generally, it is understood that in Organic farming, the products are grown under a system of agriculture without the use of chemical fertilizers and pesticides with an environmentally and socially responsible approach. This is a method of farming that works at grass root level preserving the reproductive and regenerative capacity of the soil, good plant nutrition, and sound soil management, produces nutritious food rich in vitality which has resistance to diseases. (APEDA)

Social impact assessment is a useful tool to understand the potential range of impacts of a proposed change, and the likely responses of those impacted in the event of the change. Although the scope of economics is broad, agricultural economics tends to focus on maximizing yields and efficiency whereas the social impact assessment tends to underplay the economic parameters. Socio-economic Impact assessment recognizes the complementary nature of social impact assessment and economic impact assessment. An integration of these two approaches can provide a cost effective and comprehensive method to analyse and appraise the impact of any intervention or change in a fuller manner.

WHY SOCIO-ECONOMIC IMPACT ANALYSIS?

Organic farming, as a way of farming has its share of supporters and detractors. The judgments made about the impacts of organic farming are generally found to be biased either way. The true impact of organic farming is not a subject matter for such biased judgments. Socio-economic Impact analysis provides us with an efficient and sufficient tool for making unbiased, scientific assessment of the true impact of organic farming. A social scientist has to study the entire process of organic farming and effects of organic farming in terms of human consequences, social change , social costs, opportunity costs, externalities(both negative and positive),economies of scale etc.

The inter linkages between social and economic considerations in judging the worth or value of interventions or technologies is now widely recognized .The impact of any agricultural farming system is multifaceted and one-dimensional measures , either in terms of biological productivity, net changes in income accrued to farmers , changes in input usage etc fail to capture the impact in totality. The research objectives have also, progressively moved from simple and well defined entities to objectives of complex and multiple dimensions. The analysis of impacts of any change has thus concurrently become increasingly complex. Social, Economic, Technical, technological, Political impacts studied in isolation cannot provide a comprehensive picture of the real impact of a change. The Socio economic Impact analysis, by placing its emphasis on the human consequences of a change in a holistic manner, successfully captures the key aspects of the socio economic dimensions of the change, which is beyond the individual methods of Impact analysis. Organic farming as defined earlier fits the description of a change which has got multiple dimensions and multiple objectives .The agriculture sector is inherently integrated with other sectors of the economy through forward and backward linkages. Any changes in the agriculture sector can be expected to hold consequences for other interlinked sectors as well. The adoption or non adoption of organic farming is one such change which has got implications for inter-sectoral linkages, both at micro and macro levels. The Impact assessment, in such a scenario demands that appropriate methods be used that can capture the complex and diverse nature of the cost and benefits adequately. Since Socio-economic Impact Assessment can quantify the changes in a wide range of socio economic phenomena and involve different perspectives, It is appropriate that the tools of socio- economic Impact assessment be used to measure the impact of organic farming .

LEVELS OF IMPACT ASSESSMENT

Socio-economic Impact Assessment is concerned with assessing the type, extent and duration of outcomes of any change on communities, groups and individuals. This method of impact assessment studies the equitable justice in distribution of benefits and cost through an inter disciplinary approach involving diverse perspectives. The technique is also useful for quantifying the negative and positive social and economic outcomes, and helps to identify suitable interventions to enhance the benefits and to mitigate socio economic cost. The scope and nature of impact assessment depends on the level of impact assessment that an investigator would like to conduct. The methodologies adopted, the indicators used for measuring the impact and criteria for each chosen indicator vary with the level of impact assessment. The direction of movement, the magnitude of change and qualitative changes if any, in the indicator variables are equally important in measurement of impact. Basically three levels of impact assessment of organic agriculture are followed. The choice among these levels depends on the purpose of the investigation, availability of data, availability of trained manpower, financial resources at our disposal and the time available for conducting the study.

Farm /Household / Individual level Impact—This approach uses micro level studies of the individual farm house holds to determine the impact of organic farming. The well being of the farmer, farm business income ,profit from farming , House-hold food security , risk variability, intra household gender analysis etc are used to draw conclusions on the impact of organic farming. At the individual level changes in nutrition, health, socio economic status, , employment etc are also a part of such kind of studies.

Regional level or Sectoral Level impact—The impact of Organic farming at the regional level is best reflected on the performance of the agriculture sector of the region under consideration. A comparative study of agricultural growth at the regional level brings out the impact of organic farming. At the sectoral level aggregates of socio economic variables are considered to be good indicators that help in making informed judgments about impacts. These indicators include production, farm gate prices received by the farmers, employment, income and its distribution, value added, value of output per unit area of arable land, nature of employment generated ,total employment generation , inter regional trade , poverty, labour absorption, migration, regional food security etc. The indicators can be either extensive in nature or it may measure the intensity of an indicator variable. Regional Impact Analysis can be used to identify the socio-economic significance of organic farming to a well defined regional economy .The changes in resource allocation pattern within the region can also be understood from such analysis. The cross sectoral linkages of organic farming with other sectors to identify and quantify the causal chain of policy impact and the magnitude of likely policy changes with respect to organic farming.

Economy Level Impacts—Of the various levels of impact assessment, economy level impacts of organic farming are the most comprehensive in nature. But the economy level assessment is concerned with macro level aggregates and is more popular with policy planners and stake holders than at the community level direct stakeholders. The changes and impacts of organic farming on National income, saving nature and rate of savings, trade, poverty and food security are viewed at the macro level to draw conclusions about the net impact of organic farming. Since certified organic farming in India occupies only a small proportion of cultivated area and due to the geographically dispersed nature of these areas a national economy level socio economic impact measurement at this point of time may not be feasible. For economy level assessment, the share of organic farming area in total area or the value of organic output in total value of agricultural output should be of significant size.

IMPACT DOMAINS OF ORGANIC FARMING

The impact of organic farming can be studied from different perspectives. For example, the technical impact perspective of organic farming relates to the change in the yield levels and technological inputs used in organic farming. The socio economic impact assessment includes the study of impact of organic agriculture on diverse socio economic variables. Some of these socio economic variables are discussed below.

Income—The changes in income and the direction of changes experienced by producers and consumers form the core of study of this indicator variable .The impact of organic farming on the farm income of the farmer can be either positive or negative. The impact is transmitted through the changes in production, productivity and the price of the organic produce realized in the market. If the production and productivity rise with the adoption of organic farming, it may result in an excess supply of the produce at the regional level leading to a decline in prices of the commodity. The consumers of the produce will be benefited from such higher food availability and lower prices. The cost of cultivation under organic farming is another important factor which effectively determines the farm income of the household. It can be seen that there is no fixed pattern in the relationship between adoption of organic farming and cost of cultivation. Organic farming entails cost reductions from avoidance of costly inputs like chemical fertilizers and pesticides, but it also calls for adoption of certain agronomic practices and pest management strategies which involves additional cost through increased labour usage or though the increased use of certain inputs permitted under organic farming.

Employment and labour demand—Productive employment lies at the root of all developmental efforts. The extent of socio economic impact of organic farming depend on its comparative advantage or disadvantage over conventional farming in its ability to provide adequate employment, primarily in the rural areas where employment generation is the key for overall development. The labor absorption capacity and employment generation through adoption of organic farming is relatively easier for measurement and hence an important indicator of socio economic impact. Apart from the absolute effect by way of new employment generated, issues such as loss of employment avoided, the qualitative changes in the labour requirements, skills and environment, the shift in the nature of the employment created within the community and region can be studied to get a wider understanding of the impact of organic farming on employment. Research on the employment impact of organic farming typically indicates a positive impact. The magnitude of this impact will further be dependent on the type of crop or livestock and the extend of forward and backward market linkages for the produce. There are certain crop enterprises which require less of labour under organic farming than under the conventional cultivation practices. The organic production of forage crops is an example for this kind of situation. For a country like India , where the marginal productivity of agricultural labour in rural areas is argued to be negative or near zero, the implications of labour absorption and employment generation through organic farming holds significant lessons. Much of the research focus to date has been on the quantitative aspects of employment generation while qualitative issues have received little attention. One argument is that the additional demand for labour employment created through the convertion to organic farming is largely part time employment and casual labour. Coupled with the inherent seasonal nature of labour demand in agriculture, this shift in nature of labour demanded can exacerbate the problems of seasonal unemployment and labour migration. These changes in the qualitative aspect of labour have got social significance in terms of livelihood security and employment sustainability. In India, this argument, though valid does not compound the problem of casualisation of labour since organized employment in agriculture sector is at very low levels in the existing scenario. Any additional employment opportunities should be seen as an improvement under this condition. Effective ways and means of addressing the problems of casual labour including job security needs to be explored. The gender distribution of the workload under organic farming practices and its comparison with traditional farming provides an useful tool to analyse the gender impacts, which has wider consequence with regard to social impact organic farming practice.

SKILLS, KNOWLEDGE AND NETWORKS

Practice of organic farming in India has been characterized by use of a large reservoir of indigenous technical knowledge and its combination with modern agricultural research outputs applicable for organic farming. In vast tracts of the country, organic farming is rather a way of life rather than a question of preference or choice. The skills and knowledge requirement in converting a conventional farm to organic farm and maintaining an organic farming practice are different. In the Indian context, the potential of conversion to organic farming should be studied along with the scope for realizing premium prices for uncertified inherently organic produce. These are two different dimensions of the problem. The maintenance of yield levels and control of pest and diseases under organic farming practices are knowledge intensive. The solutions to these problems are not always available of the shelf. The skills, knowledge and kinds of external linkages of innovative farmers become important under this situation. The adoption of organic farming requires that the members of the community actively seek and implement solutions in an innovative manner. The combination of Indigenous knowledge and new information require a unique set of skill. The diffusion of the successful practices also require strong network among the practicing members of organic farming. While the simple innovation diffusion model has been subject to considerable academic criticism. It nevertheless highlights a range of factors concerning organic farming and organic farmers that may help contribute to rural development (Padel 2001).Organic farming is essentially an innovation and the innovators tend to have higher educational levels and contacts and information sources from outside the community. Strong social networks within the community influences the spread and adoption of organic farming and the strength of social networks are in turn strengthened by the continuous flow of information through the social networks which are essential for the successful adoption of organic farming. The information regarding markets, cultivation practices, prices and product requirements are more intensive and extensive in organic farming and strongly influence the probability of sustainable adoption of organic farming practice. Organic farming is an information intensive innovation with the innovators and subsequent adopters actively seeking sources of information about organic farming outside of the mainstream of agriculture and from others involved in organic farming. In this context the development of skills, knowledge and social networks of the farmers who adopt organic farming can be seen both as a cause and consequence of the practice. This impact of organic farming has implications for the socio-economic status of the farmer directly and indirectly.

The networks play an important role in diffusion of a technology or change in technology. Networks are the mechanisms by which information are brought to the potential adopters or practicing organic farmers from trusted sources. The degree of trust placed on the information varies with the source of information. This information can come from personal networks or from other networks including institutional networks. Organic farming, being knowledge intensive has the potential to strengthen the networks and this holds significant benefits for farmers and farm business in that these networks will also impact the ability of the farmer to respond quickly to market signals leading to overall economic efficiency.

SOCIAL CAPITAL AND COHESION

The socio-economic impact of organic farming is reflected on essentially social parameters like social capital and social cohesion also. The sum total of all the social resources an individual commands or can potentially avail constitutes his social capital. Social capital is quantifiable and is dependent on quantitative and qualitative aspects of social networks. Organic farming involves development of more social networks and linkages since it is a knowledge driven innovation. At the community level, wide spread adoption will lead to better linkages between different constituents and may result in changes in social capital. The economic performance of any unit cannot be studied in isolation; rather it is intimately linked to the social networks and social capital of the system within which it operates. Secondary impacts such as lowering of transaction cost can be studied as a part of these changes in social capital and related changes in the community. The tendency of the organic farmers for geographical aggregation acquires importance in this context. The aggregation of organic farms may have economic impacts such as lower cost of storage infrastructure, lower cost of compliance, greater marketing efficiency etc. Geographical aggregation also benefits in forming networks which reduces the cost of information and skill sharing.

NON SOCIO-ECONOMIC IMPACT DOMAINS

The impact domains so far discussed are all related to the socio-economic impact of organic farming. The discussion on impact domains would remain incomplete without a mention of non socio-economic impact domains of impact of organic farming which directly or indirectly influence direction and magnitude of socio-economic variables. The major non-socioeconomic domains of impact include the impact on biodiversity (both at farm level and off farm level), impact on farm diversification, and impact on the supply of environmental goods and services. Adoption of organic farming leads to diversification of activities. The reason for this may be seen in the linked nature of enterprises under organic management. In conventional farming, the enterprises are independent and any complementarity arising in their operation is more of a opportunistic phenomenon than a planned one. But, in case of organic farming, there exist significant reductions in cost of production in presence of diversification. The non socio-economic impact domains are also important in impact assessment as the impact on these factors can alter the net social value of a change in a significant manner.

STEPS IN SOCIO-ECONOMIC IMPACT ASSESSMENT

Flexibility in usage of appropriate tools and methodologies and case to case variation in steps of socio-economic analysis notwithstanding, a frame work of basic steps in socio-economic impact assessment have been followed in most of the impact assessment studies. The selection of the framework for socio-economic impact assessment and subsequently, the steps to be followed depends on the specific situation and the purpose of impact assessment. However a general case of the steps followed in impact assessment is discussed here. This can be followed in socio-economic impact assessment of organic farming.

1. Definition of the Nature and boundaries of Impact Assessment

The precise target domain of the proposed socio-economic impact assessment is defined in this step. We find answers to the questions like which technology or change the impact assessment will study? At what point of time of the technology diffusion period will the study be conducted? The end users of the results of the study will be identified and the objectives of the study will be clearly spelt out in unambiguous terms. These shall act as the terms of reference for the researcher in guiding his further actions. In short the scope of the study is clearly defined and agreed upon by the users of the study and the personnel who will be conducting the study.

2. Identification of the level of Impact Assessment

The determination of the level of impact assessment is made at this stage . The level of the impact assessment influences the further choice of methods adopted and the tools to be used in the conduct of the study. The level of socio-economic impact assessment of organic farming can be at individual level, community level, sectoral level, regional level or even economy or national level.

3. Determining the Type and Feasibility of Impact Assessment Needed

After the level of impact assessment is determined, the type of impact assessment required should be chosen. In our case we have purposively selected socio-economic impact assessment for impact assessment. At this point the researcher can consider whether to employ other types of impact assessment (Environmental impact assessment, Social Impact assessment, technological impact assessment etc.). The feasibility of the impact assessment study is judged from applying many considerations like the possibility of impact assessment yielding reliable and valid information, the presence of conflict of interest, the effect of impact assessment on the welfare of the participant stakeholders, the monetary and human resources at our disposal etc. The time available for conducting the impact assessment and the commitment to use the results are also important to be evaluated at this stage

4. Profiling the Base Line Scenario

The results of the impact of any change have to be compared with some accepted standard. The base line scenario or the situation which exists in the absence of the intervention under study is used for this purpose. Though comparative studies between different interventions can also be done, comparing with a base line scenario is the primary method adopted in impact assessment study. Profiling the base line scenario provides the impact assessment, the ability to measure the incremental impact of the intervention or change. This also starts as a point for identification of key indicators relevant to the study and also to understand the range and scope of changes that can be envisages as a result of adoption of the technology.

5. Identification of Stakeholders

The success of the impact assessment depends on fruitful interaction with the people affected by the change, the impact of which we are measuring. This process requires the identification of the stakeholders and getting their involvement in the process of socio-economic impact assessment. The participation of all groups of stakeholders in the impact assessment process protects the study from personal and professional bias. The stake holders have to be consulted in developing the key questions to be answered from the impact assessment study. A thorough understanding of the indicators to be used in the study should be developed from the consultations with the stakeholders. Since the stakeholders are most intimately related to the project, their knowledge of the outcomes of the change in technology or intervention should be made use of in development of indicators.

6. Identification of Benefits, Cost and Key Indicators

The next step in the process of impact assessment is the identification of benefits and cost of the change. This requires both conceptual abilities and practical knowledge of the problem at hand. The benefits generated need to be classified into appropriate groups like economic, political, social and technological benefits, private and public benefits, tangible and intangible benefits etc. The categorization of benefits into physical, financial and human resource benefits is another scheme which is usually followed in impact assessment. The benefits and cost should also be classified according to the groups to which they accrue. This will help in judging the distributional equity of benefit and cost sharing. An in-depth knowledge of the subject matter under study, organic farming in our case, will be of help in visualizing the potential impacts and the associated benefits and cost from those impacts. In impact assessment, the benefits and cost generated as a result of adoption of organic farming are measured using indicators. The choice of indicator is critical to the success of impact assessment. The indicators used should have specificity, reliability, validity and should be easily measureable.

7. Identification of Approaches and Techniques for Data Collection

With the knowledge of the indicators to be used and the requirement of qualitative and quantitative data for the study, the approach to be adopted for data collection can now be determined. The techniques of data collection vary from personal interviews, focus group discussions. The sampling framework, sample size and sampling method is selected which suits the data requirements of the impact assessment study.

8. Collection, Analysis and Interpretation of Data

The actual data collection, its tabulation and analysis forms the next step of impact assessment. Proper statistical tools are applied to the collected data at this stage and valid conclusions are drawn from the data collected from the sample.

9. Estimation of Impacts

The stage is now set for estimation of the impacts of the organic farming. The impacts may include the ex-post impacts of the change and the anticipated changes at a future point of reference (ex-ante). The impacts are estimated with respect to their magnitude and direction of impact. The previous nine steps in impact assessment gives us the necessary inputs to make this estimate specific and relevant to the change considered in the study. Quantifying the impacts and identifying the population affected can help in measuring the net socio-economic impact of the change. Once the estimates of impact are obtained, informed value judgments can be made about the desirability or otherwise of the studied change or technology.

10. Communication and Application of Findings

The findings of the impact assessment study are communicated and discussed with the users of the information including the stakeholders at different levels. Along with the communication of the socio-economic impact assessment study, mitigation and management strategies for the identified impact cost of organic farming has to be discussed as potentially viable alternative strategies. The users of the impact assessment study may vary from the community directly affected by the adoption of organic farming to the policy makers at government level. They may also include non-governmental stakeholder agencies and various developmental institutions. The communication of the results of the impact assessment should be unbiased and constructive to the extent possible.

11. Monitoring and Evaluation of Impact Assessment Programme

The impact assessment process does not end with the communication of the results of the study. A continuous process of monitoring and evaluation of the impact assessment process itself need to be undertaken to identify the shortfalls in the process adopted for socio-economic impact assessment and to rectify them . The focus of mitigation of negative externalities should be to enhance the benefits, minimize the cost and to compensate adequately for the cost borne by the members, especially in cases where cost disproportionate to the benefits have to borne by a section of the stakeholders. The evaluation should enhance the reliability and validity of the study through process control and course corrections.

ISSUES IN SOCIO ECONOMIC IMPACT ASSESSMENT OF ORGANIC FARMING

The socio-economic impact assessment of organic farming uses tools and techniques derived from interdisciplinary sources. For a social scientist, impact assessment generates certain critical questions and issues which need to be addressed adequately. The issues involve methodology used, tools employed or approach adopted. We often find two mutually exclusive and divergent views at two ends of the continuum of a particular issue. The consensus about the resolution of the problem usually involves a synthesis of the two mutually exclusive views yielding a composite tool, method or approach. Some such issues are discussed below.

The balance in the use of qualitative and quantitative data is one such issue. In many impact assessment studies there is a marked overdependence on quantitative data. The over emphasis on quantitative data can cloud the real social impact of a change since many of the changes or impacts resulting from organic farming are qualitative in nature. Quantitative data lends easiness in comparison of magnitude of impact between different levels or regions. Where ever appropriate quantitatively measureable indicators are available, they provide conceptually simple measurement tools for socio economic impact assessment. The situational changes revealed through hard statistics is easily understood by a wide range of stakeholders. The qualitative data on the other hand calls for specific skills and conceptual ability for proper measurement. Many of the changes in socio-economic parameters require use of qualitative variables to adequately capture the impact of organic farming. The changes in community networks, changes in social capital etc are some impact areas where the use of qualitative variables is important for proper measurement. Qualitative variables study the nature of the indicator variable whereas the quantitative variables study the extend. A comprehensive approach involving the use of both qualitative and quantitative data is usually followed in socio –economic impact assessment.

Along with the impact assessment, it would be considered important to identify the drivers of impact and the transmission mechanism of the impact. The socio-economic impact of organic farming can be viewed from the perspective of economic drivers of impact or the social drivers of impact. The selection of approach shall depend on the purpose of assessment. The economic transmission of impact happens mainly through changes in supply and demand for inputs and outputs. The direction of technical change and the rate of technical change, which are key to the sustainability of organic farming, are also influenced by the supply and demand factors. Measurement of rate of technical change in organic farming systems and its decomposition into constituent factors at micro and macro levels can throw light on many important issues like identification of sources of technical change, relative importance of the change agents at various levels and policy implications of diverse sources of technical change.

Impact assessment usually places its focus on long term impacts or changes. A key point of debate is the time frame after which an impact assessment can yield the true picture of impact of organic farming. There are classes of impact indicators which takes time to change in response to the adoption of organic farming. For example, social indicator variables of impact like changes in community networks and social capital takes time to show significant changes. The impact assessment after a short period of time, say within two years may not be adequate to capture such changes. In contrast there are certain impact indicators which respond in the short run itself. Many of the economic variables fall under this category. The selection of time frame for impact assessment should take in view these subtle issues which can affect the results of impact assessment. The review of the results should keep in mind the time frame used for the study in judging the results.

The benefits or cost of changes resulting from organic farming at all the levels fall into two broad categories viz, tangible and intangible. The measurement of tangible changes tends to be easier, so there exist a propensity to ignore the intangible benefits and cost of socio –economic impacts of organic farming. This bias against intangible attributes of change often leads to erroneous conclusion regarding the net socio-economic impact. Adequate care must be taken during all steps of impact assessment to include indicator variables to measure these intangible costs and benefits. Change in individual satisfaction from farming, perceived social cohesion, trust, etc. are a few examples where intangible attributes are involved. Suitable indicators to capture these changes should be developed and incorporated in the Impact assessment study.

The level of participation of the stakeholder individual or community members in impact assessment is a point of debate among social scientist. The recent trends in impact assessment lay stress on participatory approaches to impact assessment. The underlying philosophy of this approach is that the stakeholders are able to precisely identify all the impact domains of the change. Participatory and non-participatory approaches to impact assessment are different in methodology and usage of tools of impact assessment. Stakeholder participation in the impact assessment process enhances the reliability and validity of the study. Non participatory methods often rely heavily on outside expertise in conduct of impact assessment and may fail to provide quick feed back to the stakeholders. The lack of intimate knowledge of community dynamics also works against a comprehensive understanding of the socio-economic impact. The impartial and unprejudiced judgment of the impacts is one of the advantages in having non participatory impact assessment. Depending on the variables under study a combination of both participatory and non participatory methods will have to be employed for impact assessment of organic farming.

The distributional aspects of benefits and cost among the members of the stakeholder community are another area of inconclusive debate. The issue here is whether to look at the net social benefits alone or to be concerned with the distributional justice of the benefits and cost. The ideal scenario would be that the benefits and cost arising out of adoption of organic farming be equally shared by the stakeholders. But, this is seldom found to be the case in reality. This raises subsidiary issues such as the compensation for negative consequences and equitable sharing of benefits.

EMERGING PERSPECTIVES IN SOCIO-ECONOMIC IMPACT ASSESSMENT

The area of impact assessment has over the years developed substantially drawing on the experiences from the empirical studies conducted in this field. The complex nature of impact assessment and its utility to make informed welfare policy decisions is now beyond question. The recognition of non-economic factors as a key element of impact assessment was a major catharsis in the field of impact assessment. It is now widely recognized that impact of a change or technology does not occur in isolation. It influences and is influenced by the back drop of existing policy environment and the level of technology. The key role played by institutions in the society, which was overlooked in earlier studies is receiving more and more attention. The institutional impacts of the change can be as important as any other kind of change a institutions form the basic building block of organized economic activity. The changes in infrastructure and level of infrastructure already available within the target domain are also of importance in impact assessment. How these factors change often dictate the magnitude and direction of other socio-economic impact domains. Policy, technology, institutions and infrastructure form a matrix of interaction which defines the space and scope for socio-economic changes due to a particular technology or change.

The relative role played by private and public agencies in acting as agents of change in adoption of organic farming is crucial to the nature of diffusion of the technology and in determining the distributional aspects of benefits and cost. Sustainability of the changes brought about by the adoption of organic farming is gaining importance in view of the fact that many of these changes are reversible in nature and in-fact tend to do so. Some changes are dependent on policy- technology-institution-infrastructure matrix and changes in these factors could change the dimension of the impact. For example, with large scale adoption of organic farming and the process of incremental innovation, technological change could happen in such a way as reduce the labour absorption in organic farming. Such scenarios are typically not incorporated in the impact assessment studies. A long run appraisal of organic farming impact should give adequate space for such considerations.

The area of socio-economic impact assessment is an evolving field of study. There is no single method of study that can claim to be the best. The choice of method, tools an approach depend on a variety of factors. The key to successful conduct of socio-economic impact study lies in having a proper understanding of interdisciplinary nature of the problem. If done in a scientific manner, socio-economic impact assessment can provide reliable, valid, useful and actionable inputs to the policy planners.

REFERENCES

Granovetter, M. 1985. "Economic action and social structure: the problem of embeddness." *American Journal of Sociology* 91: 481-510.

Midmore, M. and N. Lampkin 1994. Modelling the impact of widespread conversion to organic farming: an overview. In The Economics of Organic Farming: An International Perspective. N. Lampkin and S. Padel (eds). Wallingford, *CAB International*: 371-380.

Padel, S. and N. Lampkin 1994. Conversion to Organic Farming: An Overview. In The Economics of Organic Farming: An International Perspective. N. Lampkin and S. Padel (eds). Wallingford, *CAB International*: 295-313.

www.fao.org

www.ifoam.org

❑❑❑

Chapter 19

Organic Farming and Food in India

Joyti Bajpai* and Sanjay Ranjan**

**Research Associate, IPR Central Institute for Subtropical Horticulture (ICAR) Institute Rehmankhera, Lucknow*

***Senior Research Fellow, National Botanical Research Institute (CSIR), Lucknow*

INTRODUCTION

Definitions

Organic agriculture represents a broad set of practices that emphasizes farming based on ecosystem management, integrated cropping and livestock systems, diversity of products and reliance on natural pest and disease control without the use of synthetic inputs. The objectives of organic agriculture are to produce sustainable and healthy food through harnessing natural biological and ecological processes. The Codex Alimentarius Commission (2001) defines organic agriculture as "a holistic production management system which promotes and enhances agro-ecosystems health including biodiversity, biological cycles, and soil biological activity. It emphasizes the use of management practices in preference to the use of off-farm inputs, taking into account that regional conditions require locally adapted systems. This is accomplished by using, where possible, cultural, biological, and mechanical methods, as opposed to using synthetic materials, to fulfill any specific function within the system". An organic agricultural system is therefore designed to rely on renewable energy sourced as much as possible from on-farm or natural local systems. Although organic agriculture adheres to certifiable standards, farmers have the flexibility to enhance the ecological and sustainable practices of their farms beyond what the standards require and **Organic farming** is a production system of crops which avoids the use of synthetic and chemical inputs like fertilizers, pesticides, growth regulators and livestock feed additives. Organic farming systems depend mainly on crop rotation, crop residues, animal manures, legumes, green manures, off-farm organic residues, mineral bearing rocks, natural pesticides and biological pest control to maintain soil productivity and supply important nutrients like N, P, K. Indiscriminate use of chemical fertilizers for decades has resulted in the decline of organic matter content of soils to less than one per cent. The use of pesticides has led to pest resurgence and difficulty in controlling weeds. Indiscriminate use of artificial fertilizer and pesticide has caused severe water and environmental pollution. The residue of chemical is of serious concern for the safety of food and sustainable production. Organic farming ensures the production of palatable food free from chemical residues.

INDIA'S FUTURE LIES IN ORGANIC FARMING

Dilemmas dominate the life of Indian farmers. They have immense capacity to endure hardships - natural or man-made. Their farming wisdom remains unchallenged despite technological advancements. The Green Revolution was the finest outcome of their vigorous performance even during crises but the frequent policy changes have always affected their performance. Inconsistent policies create dilemmas, which sometimes prove disastrous. Now, the farmers are being persuaded to switch over to the age-old organic farming and thus phase out the consumption of chemical fertilizers. It is again a policy-driven dilemma. No doubt that organic food is the best for human health but it is not so easy for the performance-oriented farm sector to revive a conventional practice until it becomes sensitive to the ecological crises ahead.

Organic farming is still in practice in many parts of the country. The farming community had been using bio-fertilizers until using chemical fertilizers. Repeated warnings about the harmful use of chemical fertilizers were grossly ignored in the euphoria of the Green Revolution. The incessant use of chemicals eroded the soil fertility and polluted the ground water reserve. Increase in fertilizer consumption has significantly contributed to the sustainable production of food grains, many health problems have surfaced in the last three decades. It was reported that the NPK consumption has also increased from 0.7 lakh to 167 lakh million tonnes since 1951-52.

WHY ORGANIC FARMING?

Every year 5.33 million tonnes of soil erodes in India and with it 53.3 lakh tonnes of NPK flows away according to report. The three major components of bio-farming are integrated plant protection, integrated pest control and soil and water management. A task force on organic farming set up by the Government of India defines bio farming as a holistic production management system, which promotes an agro eco-system of health including bio-diversity, biological activity. Evolution of the concept of farming has been best illustrated in Kautilya's Arthashastra, Panini's Astadhyayi, Patanjali's Mahabhashya and Varahamihira's Brihat Samhita. The sage Parasara in Krishi Parashara in the 10th century stated that the "life of farmers is solely dependent upon the microbes present in the soil." The approach and outlook towards agriculture and marketing of food has seen a quantum change worldwide over the last few decades. Whereas earlier the seasons and the climate of an area determined what would be grown and when, today it is the "market" that determines what it wants and what should be grown. The focus is now more on quantity and "outer" quality (appearance) rather than intrinsic or nutritional quality, also called "vitality". Pesticide and other chemical residues in food and an overall reduced quality of food which have led to a marked increase in various diseases, mainly various forms of cancer and reduced bodily immunity. This immense commercialization of agriculture has also had a very negative effect on the environment. The use of pesticides has led to enormous levels of chemical buildup in our environment, in soil, water, air, in animals and even in our own bodies. Fertilizers have a short-term effect on productivity but a longer-term negative effect on the environment where they remain for years after leaching and running off, contaminating ground water and water bodies. The use of hybrid seeds and the practice of monoculture have led to a severe threat to local and indigenous varieties, whose germplasm can be lost for ever. All this for "productivity"!

PESTICIDES YOU COULD FIND IN YOUR FOOD

The twin controversies in 2003 regarding pesticide content in bottled drinking water and aerated beverages in India hardly came as a surprise to many working with the environment and in farming. The pesticide problem is compounded in India because many pesticides banned abroad are manufactured / dumped and sold freely here. Pesticides are not bio-degradable, are highly toxic and find their way into ground water and water bodies, contaminating them and rendering them unfit for drinking purposes. Remember that even if you blame (though rightly-so) a beverage manufacturer for allowing pesticide residues in their products and treating human life so cheaply, the fact remains that pesticides entered the water supply in the first place only because of the agriculture system which used them.

These are some of the pesticides present in the food we eat.

Table 1 : Food product contaminant pesticides

Fruit/food products	Chemical contamination
Apples	Diphenylamine, Captan, Endosulfan, Phosmet, Azinphos-methyl
Bananas	Diazinon, Thiabendazone, Carbaryl
Cabbage	Methamidophos, Dimethoate, Fenvalerate, Permethrin, BHC
Carrots	DDT, Trifluralin, Parathion, Diazinon, Dieldrin
Cauliflower	Methamidophos, Endosulfan, Dimethoate, Chlorothalonil Diazion
Grapes	Captan, Dimethoate, Dicloran, Carbaryl, Iprodione
Onions	DCPA, DDT, Ethion, Diazinon, Malathion
Potatoes	DDT, Chlorpropham, Dieldrin, Aldicarb, Chlordane
Tomatoes	Methamidophos, Chlorpyrifos, Chlorothalonilo, Permethrin, Dimethoate

Source: "Return to the Good Earth", Third World Network

This is where organic farming comes in. Organic farming has the capability to take care of each of these problems. Besides the obvious immediate and positive effects organic or natural farming has on the environment and quality of food, it also greatly helps a farmer to become self-sufficient in his requirements for agro-inputs and reduce his costs.

TECHNIQUES, PRACTICES AND SPECIALIZED FORMS OF ORGANIC FARMING

Some of the techniques and practices integral to organic farming are detailed below. This is not complete in itself, and farmers can innovate and adapt to suit their surroundings. That is what organic farming is: the answers to a problem should come from the farmer, his fields and his surroundings rather than from a chemical factory or the village pesticide shop. Moreover, there cannot be a fixed package of practices - every area is unique in its own way, has its own endemic species - both plant and animal - and its own natural conditions.

Biodynamics

Biodynamics is essentially a science of life forces and recognition of the role of nature and "higher forces" in agriculture.

In Biodynamics, the earth is considered a living being in a living universe, part of a spiritual-physical matrix. All substances are carriers of life-creating forces and interact with celestial rhythms, which directly affect life. All living beings therefore are manifestations of celestial rhythms. A farm is considered a living, dynamic and spiritual entity with its own rhythms and life forces. Biodynamics is a vast subject and a farm can be managed entirely on its principles and practices. It has two basic components - farming operations on the basis of an astronomical calendar, and the use of some very special preparations, which are used as sprays and in the compost heap. Biodynamics addresses soil and plant health, pest control, composting, the relationship of the soil and the plant, food quality, animal husbandry and animal welfare measures. It recognizes that there is an abstract but intrinsic quality of food (vitality) that cannot be measured by a scientist calculating nutrient percentages in a laboratory. Biodynamics is today perhaps the largest "specialised" organic farming system in the world. Farms certified biodynamic can market their produce under the "Demeter" symbol, which is recognised by the BD community worldwide - consumers, buyers, traders and farmers.

Composting

A huge quantity of crop wastes/residues and animal wastes are always available on a farm. The common practice is to burn plant wastes, which, besides being an environmental disaster, is also a waste of the huge potential of these residues. Properly recycled, these residues form excellent compost in one to six months, depending upon the composting process used. Every farm can choose or even develop a suitable compost process depending upon its own needs and resources, including availability of labour, managerial time and investment potential.

Mulching, Green Manuring and Cover Cropping

All these techniques are different but somewhat interrelated. Mulching is the use of organic materials (plastic mulch is expensive and non-biodegradable) to cover the soil, especially around plants to keep down evaporation and water loss, besides adding valuable nutrients to the soil as they decompose. Mulching is a regular process and does require some labour and plenty of organic material, but has excellent effects, including encouraging the growth of soil fauna such as earthworms, preventing soil erosion to some extent and weed control. Green manuring is an age-old practice prevalent since ancient times. A crop like dhaincha (Sesbania aculeata), sunnhemp or horsebean is sown (usually) just before the monsoons. A mix is also possible. Just around flowering (30-45 days after sowing), the crop is cut down and mixed into the soil after which the season's main crop is sown. Green manuring is beneficial in two ways - firstly it fixes nitrogen, and secondly the addition of biomass (around five to ten tons/acre) greatly helps in improving the soil texture and water holding capacity. Green leaf manuring can also be carried out if sufficient leguminous tree leaves are available. Cover cropping is normally carried out also with nitrogen-fixing crops that grow fast and require little or no inputs like water or additional manuring. While cover crops can yield some returns, they are mostly used for covering the soil in the fallow months, adding nitrogen to the soil, suppressing weeds, preventing soil erosion and later used as biomass or fodder. Velvet bean is an example, and it finds use as a fodder crop and biomass generator. Another useful cover crop is Dolichos lablab, which is a source of fodder and food.

Crop Rotation and Polyculture

One of the most important aspects of organic farming is the strict avoidance of monoculture, whether annuals or perennials. Besides the proverbial "putting all eggs into one basket", monoculture systems are unhealthy for the ecosystem they are a part of the prime requirement for any natural ecosystem to thrive and be healthy is diversity. Traditional farmers till date follow the systems of crop rotation, multi-cropping, inter-cropping and polyculture to make maximum use of all inputs available to them, including soil, water and light, at a minimum cost to the environment. The home gardens of Kerala are an excellent example.

Crop rotation—Crop rotation is the sequence of cropping where two dissimilar type of crops follow each other - a few examples include cereals and legumes, deep-rooted and short-rooted plants and where the second crop can make use of the manuring or irrigation provided some months earlier to the first crop (eg. rice + wheat, rice + cotton). The combinations possible are endless, and will depend to a great deal on the local situations.

Multi-cropping—Multi-cropping is the simultaneous cultivation of two or more crops. In Indian agricultural tradition, farmers have been known to sow as many as 15 types of crops at one time. An example of multi-cropping is Tomatoes + Onions + Marigold (where the marigolds repel some of tomato's pests).

Inter-cropping—Inter-cropping is the cultivation of another crop in the spaces available between the main crop. A good example is the multi-tier system of coconut + banana + pineapple/ginger/leguminous fodder/medicinal or aromatic plants. While ensuring bio-diversity within a farm, inter-cropping also allows for maximum use of resources. All these are forms of polyculture and biodiversity and help in keeping pest populations in control. Fallen leaves and other crop residues in combination add more value to the soil or compost heap they become a part of, again because of the nutritional mix.

Effective Microorganisms

As the name suggests, it makes use of microorganisms, mainly lactic acid bacteria, photosynthetic bacteria, yeast, filamentous fungi and ray fungi. These microorganisms are both aerobic and anaerobic and are not genetically modified. EM, like Biodynamics can be useful in many different ways on the farm, including improving soil health, as a pest repellent and prophylactic, in composting, and in animal feeds, animal health and hygiene, aquaculture, etc. Different EM cultures are used for agriculture, animal husbandry, and aquaculture.

Integration of Systems

In nature, the whole is greater than the sum of its parts and the key to the success of any natural system is diversity. Diversity adds complexity to the farm system lending it greater stability. There are economic and productivity benefits too. The concept of polyculture should not be limited to plants only but extended to cover the whole farm. This way, one system's wastes and by-products are another system's inputs, or one system is comprised of more than one component, which allows for efficient use of available resources. An example of such integration is: rice-fish/prawn systems where the fish/prawn mature in the waterlogged fields and are harvested before the water drains away (making use of available resources). They have a symbiotic relationship with the main crop in two ways - manuring and pest control. A larger and more permanent example of integration could be: annual crops + tree crops + dairy cows + honey bees. The animals and tree crops are benefited by the honey bees (pollination); crop residues and tree prunings are useful as cattle feed, green leaf manure and in composts; the dung from the cattle is useful at the bio-gas plant, after which the slurry finds use in the fields as manure and in the compost heap. With so many benefits, one almost forgets that this farm also produces food grains, fruits, vegetables, firewood, timber, milk and honey. There is no limit to the extent and diversity of integration possible on a farm, however large or small it may be. It is important to remember that nothing on the farm is waste or useless. Greater integration or diversity also calls for better management.

Living Fence

Having a living fence around the farm has multiple benefits. Besides protection from trespassers and cattle, a living fence also to provide a buffer, and with a sensible choice of plants, even some revenue. It does however take two to three years to develop. The species suitable for live fence should be thorny, inedible and non-browsable for cattle and goats, hardy and relatively maintenance-free (other than pruning / lopping), adaptable to the local conditions, fast-growing and producing something that can yield some revenue. Suitable species for a live fence include agave, jatropha, euphorbia and horsebean. A live fence should ideally be planted just before the monsoons and watered regularly after the rainy season is over to ensure optimum growth.

Microbial Biofertilizers

Microbial biofertilizers are biologically active (living or temporarily inert) inputs and contain one or more types of beneficial microorganisms such as bacteria, algae or fungi. Every microorganism - and hence each type of biofertilizer - has a specific capability and function. It would be relevant to mention that vermicompost is not a biofertilizer as is propagated by some, but merely an improved form of compost. There are broadly seven types of biofertilizers:

Rhizobia—*Rhizobia* is a group of bacteria that fixes nitrogen in association with the roots of leguminous crops. *Rhizobia* can fix 40-120 kgs of nitrogen per acre annually depending upon the crop, rhizobium species and environmental conditions. They help improve soil fertility, plant nutrition and plant growth and have no negative effect on soil or the environment. Every leguminous crop requires a specific *rhizobium* species

Azotobacter—*Azotobacter* is also a group of nitrogen-fixing bacteria but unlike *rhizobia*, they do not form root nodules or associate with leguminous crops. They are free-living nitrogen fixers and can be used for all types of upland crops but cannot survive in wetland conditions. In soils of poor fertility and organic matter, *azotobacter* needs to be regularly applied. In addition to nitrogen fixation, they also produce beneficial growth substances and beneficial antibiotics that help control root diseases.

Azospirillum— Like *azotobacter, azospirillum* species also do not form root nodules or associate with leguminous crops. They are, however not free-living and live inside plant roots where they fix nitrogen, and can be used in wetland conditions. This group of microorganisms also produces beneficial substances for plant growth, besides fixing atmospheric nitrogen. *Azospirillum* does well in soils with organic matter and moisture content, and requires a pH level of above 6.0.

Blue-green algae—Blue-green algae or cyanobacteria are free-living nitrogen-fixing photosynthetic algae that are found in wet and marshy conditions. Blue-green algae are so named for their colour but they may also be purple, brown or red. They are easily prepared on the farm but can be used only for rice cultivation when the field is flooded and do not survive in acidic soils

Azolla—*Azolla* is a free-floating water fern that fixes nitrogen in association with a specific species of cyanobacteria. *Azolla* is a renewable biofertiliser and can be mass-produced on the farm like blue-green algae. It is a good source of nitrogen and on decomposition, a source of various micronutrients as well. Its ability to multiply fast means it can stifle and control weeds in (flooded) rice fields. *Azolla* is also used as a green manure and a high-quality feed for cattle and poultry.

Phosphate-solubilising microorganisms—These are a group of bacteria and fungi capable of breaking down insoluble phosphates to make them available to crops. Their importance lies in the fact that barely a third of phosphorous in the soil is actually available to the crop as the rest is insoluble. They require sufficient organic matter in the soil to be of any great benefit.

Mycorrhiza— *Mycorrhiza* is a sweeping term for a number of species of fungi which form a symbiotic association with the plant root system. Of these, the most important in agriculture is vesicular-arbuscular *mycorrhiza* or VAM. Plants with VAM colonies are capable of higher uptakes of soil and nutrients and water. VAM strands acts as root extensions and bring up water and nutrients from lateral and vertical distances where the plant root system does not reach. A point to consider before using biofertilisers produced by commercial units is the issue of using microorganisms native to another area or region. It is possible to isolate the required species of microorganisms from a farm's soil and mass-produce them. Besides allowing a better chance of survival of the organism(s) used, this ensures that local species of microorganisms alone are used. There are a few crude (but fairly effective) as well as standard laboratory procedures for isolation and mass production of biofertilisers.

Vermicomposting

Vermicomposting is a modified and specialised method of composting - the process uses earthworms to eat and digest farm wastes and turn out high quality compost in two months or less. Vermicompost is not a biofertiliser as is touted by some, merely improved compost. Vermicompost can also be used to make compost tea. Vermicompost tea is useful as a prophylactic against pests and diseases, for pest repelling and as a foliar spray. A by-product of vermicomposting called vermi wash (which can be collected if there is a tap at the base of the vermicompost tank) also serves the same purpose.

An important point to note in case of vermicomposting but widely ignored, is to carry out proper sieving of the compost before applying it in the fields. The most efficient and widely used earthworms in vermicomposting are not indigenous and if the worms and casts find their way to the fields, they will quickly colonise and dominate the local species. Farmers can also use indigenous earthworm species, collecting them from their fields using collection baits and introducing the earthworms into heaps. In the usual way vermicomposting is practiced in India and most other places around the world, it is both labour-intensive and requires some infrastructure. As a result, while a small farm can use this method to compost most of its wastes, a larger farm often finds it expensive and difficult to compost most of its wastes through vermicomposting. We have after some experimentation developed a methodology through which these issues are resolved - labour, time, managerial inputs, infrastructure and the usage of exotic worms.

NEED AND SCOPE FOR ORGANIC FARMING IN INDIA

Recent reports on rejection of large consignments of Indian food exports by the United States and some European countries on grounds of several sanitary and phytosanitary (SPS) measures have raised a question mark on the future of the country's agricultural exports. However, as per recent reports, Indian exports in general have performed well, registering a growth of 19.18 percent and garnering forex. Agriculture and allied products export, which account for only 8.5 percent by weight in the total basket, posted a growth of 10.32 percent by fetching $4.48 billion. Exports of plantation crops like tea and coffee could fetch only $536 million, marking a decline of 9.10 percent. The message is loud and clear. We should not be complacent with a 10.32 percent growth in agriculture and allied products. These products constitute only 8.5 percent by weight in the total export basket. In dollar terms our major exports are of rice, wheat, raw tobacco, spices, cashew nut-in-shell, oil meals, sugar and molasses, fresh and processed fruits, vegetables, meat and meat preparations and marine exports. Though tea and coffee are major items for exports they are on declining trends. We need to boost exports of all other agricultural commodities. Global consumers are increasingly looking for organic food, which is considered safe and hazard free. Countries which are opting for organic foods are Austria, Belgium, Denmark, France, Germany, Ireland, Italy, Netherlands, Sweden, Norway, Czech Republic, Estonia, Latvia, Lithuania, US, Canada, Japan and Oceania countries. The global prices of organic food are more lucrative and remunerative. Are we able to exploit this growing market, despite the fact that a large part of our country is organic by default? The government and the exporters should take up this task of locating organic farms in the country and encourage them to continue with organic farming. India can develop higher SPS norms than EU if organic farming is encouraged and our agricultural exports will not face any problems in the future. Also we will be in a better position to address the health concerns of our people.

India has so far allowed only Bt cotton and not any other GM crops. It is, therefore in a better position to export its agro produces to EU and other countries, which are averse to GM foods. The total area under organic agriculture in the world is 17.16 million-hectares. The organic food movement is gaining ground in Europe and America where populations have accorded preference for organic food. The reason for the growing preference for organically grown food is simple. It reduces serious health problems. The New Scientist reports that the consumption of organic food may reduce the risk of heart attacks and cancer. Expectedly, Indian farmers would penetrate the global organic food market with a solid market network facility. The potential of organic farming is signified by the fact that the farm sector has abundant organic nutrient resources like livestock, water, crop residue, aquatic weeds, forest litter, urban, rural solid wastes and agro industries, bio-products. India offers tremendous scope for bio farming as it has local market potential for organic products. Absence of local markets for organic products in many of the Asian countries brightens India's chances for exporting organic food. It is now the responsibility and solemn duty of the leadership to encourage organic farming.

REWARDS OF ORGANIC FARMING

One of the biggest rewards of organic farming is healthy soil that is alive with beneficial organisms. This healthy soil has its natural "immune system" intact and thus can control several diseases. Several beneficial bacteria and fungi compete with, and/or prey upon, disease-causing organisms. These 'good' microorganisms function like the white blood cells in our bodies keeping the harmful bacteria and fungi that cause disease in check.

ORGANIC LABELLING AND CERTIFICATION

The practice of alternating the species or families of annual and/ or biennial crops grown on a specific field in a planned pattern or sequence so as to break weed/pest and disease cycles and to improve soil fertility and organic matter is called crop rotation. Planting the same crop in the same location time after time (mono cropping) encourages the build up of diseases and pests that plague that particular crop. Crop rotation helps in overcoming this problem without the use of any chemicals.

The growing demand for organic food has led to the imitation of a process for development of their own standards by different countries. The US and the European Union have already announced comprehensive National Organic Programmes. Japan, Canada and Australia have national standards for organic products in practice since early nineties. New Zealand, Isreal and Brazil have almost attained the equivalency status with the standards of the European Union and the USA.

Certain very large organic food markets such as the United States present particular problems for exporters because of the rigid requirements for labelling. In the US, products are accepted as organic only if they conform to US Department of Agriculture (USDA) requirements, which are fairly stringent and require which can be costly and cumbersome for exporters from poor countries.

REFERENCES

Natural Resources Management and Environment Department Food and Agriculture Organization of the United Nations

Science tech entrepreneur September 2006

Website: www.solar-aid.org

❑❑❑

Chapter 20

Diabetes : The Chronic Health Hazards of Non Organic Farming

Ashok K. Tiwari

Scientist E-II, Pharmacology Division, Indian Institute of Chemical Technology, Hyderabad-500 007

Industrialization, Chemical, Green and White-Revolutions, non organic farming and human activities have produced several non-natural pollutants. These pollutants have affected life of rich and the poor indiscriminately. For decades, toxicologists have been warning about the harmful effects of these pollutants. Time has come now to seriously look beyond the conventional understanding of diabetes pathogenesis in finding out other set of factors that contribute to the epidemic of diabetes in modern era.

According to European Action on Global Life Sciences (EAGLES) Diabetes Report-2008, eighty per cent people with diabetes in the world hail from developing countries. This number is predicted to grow by 150% in next 25 years. The International Diabetes Federation estimates that about 3.8 million people died of diabetes in 2007 and without urgent action, diabetes deaths will increase by fifty percent in the next 10 years. This estimate is more than the deaths from HIV/AIDS and nearly four times more than the deaths from malaria.

Population based surveys of seventy five communities in thirty two developing countries observed that diabetes is rare in communities in where traditional life style has been preserved. By contrast however, some Arab, Asian Indian, Chinese, and U.S. Hispanic communities that have under gone Westernization and urbanization are at high risk. In these populations, prevalence of diabetes ranges from 14-20 per cent. Indians lead the world with greatest number of diabetes (42 million) that is projected to expand up to seventy million by the year 2025 due to the compounding effect of increasing modernization coupled with adoption of sedentary life-style in urban as well as in rural India. In addition to these developments, compared with other ethnic groups around the world, Indians are known to be more insulin resistant, secrete more insulin in response to glucose, demonstrate a higher glycemic response to all foods, and develop type-2 diabetes at younger age. Therefore, European antidiabetic prescriptions require different handling in Indian framework.

Nonetheless, despite the understanding that diabetes is a life style disease caused by laziness and gluttony and availability of array of drugs and devices in 21st Century, disease is taking shape of Global Epidemic. Knowledge about this disease has been known to mankind for centuries and has been treated as disease of the Wealthy. However, in this century diabetes is not restricted to the rich people only, and instead it has spread its tentacles in every strata of the society. Infact, in the course of modernization and wealth generation to improve living standards through the process of Growing group of scientists' world over have now started believing that the real culprit for development of diabetes in mass may be the family of toxic chemicals known as persistent organic pollutants (POPs). It has been observed recently that there is a striking, monotonic and additive dose-response relationship between serum concentrations of POPs and prevalence of diabetes in the U.S. population with background environmental exposure to POPs. This National Health and Examination Survey 1999-2002, observed a strong dose-response relation between serum concentrations of six POPs [2,2',4,4',5,5'-hexacholorobiphenyl (HCB); 1,2,3,4,6,7,8-heptachlorobenzo-*p*-dioxin;1,2,3,4,6,7,8,9-octachlorodibenzo-*p*- dioxin; Oxychlordane; p,p'-dichlorodiphenyl trichloroethane; and *trans*-nonachlore] and prevalence of diabetes. The POPs belonging to organochlorine (OC), pesticides (*trans*-nonachlore and oxychlordane) or *non*-dioxins like polychlorinated biphenyls (PCBs) subclasses were strongly associated with diabetes. Interestingly, obesity did not increase the prevalence of type-2 diabetes among subjects without detectable levels of POPs in this study. Similarly, results from the National Health and Nutrition Examination Survey 1999-2002, revealed that serum concentrations of OC pesticides were strongly and positively associated with the phenomenon of insulin-resistance, a frequent pathogenic precursor of type-2 diabetes among non diabetic subjects. It is suggested, therefore, that the background environmental exposure to some POPs, especially OCs and Chlordane pesticides, may be critically involved in pathogenesis of diabetes through the pathway involving insulin resistance. Another study from National Health and Environment Effect Research Laboratory U.S.A. observed increased mortality from Ischemic Heart Disease (IHD) and type-2 diabetes mellitus in four U.S. wheat producing States and established association between increased environmental exposure to chlorophenoxy herbicides used in great wheat producing States and mortality from type-2 diabetes in general population of the county. This study supported earlier observations on effects from exposure to dioxin and dioxin-like compounds, and increased mortality from IHD and diabetes in the general rural population of agricultural counties in States like Minnesota, Montana, North and south Dakotas. Furthermore, association between increased prevalence of self-reported diabetes in people working in pesticide plants and farming occupations exposed to high level of β-HCH, HCB, p, p'-DDT and oxychlordane, were also observed.

The association between exposure to inorganic arsenic and incidence of diabetes is a relatively new toxicological observation. Up to now there are twenty epidemiologic studies linking diabetes mellitus with acute as well as chronic exposure to arsenic from environmental and occupational sources. Cross-sectional and prospective follow up studies from Taiwan indicated the association between arsenic exposure from drinking water and prevalence of diabetes in a dose-dependent manner. Further reports from Sweden and Bangladesh supported these studies. However, except the studies from Taiwan and Bangladesh, other studies could not show consistency in a systematic analysis. Recently, in a cross-sectional study from National Health and Nutrition Examination Survey (USA) 2003-2004, Ana *et al* found that after adjustment for biomarkers of seafood intake that constitute organic arsenicals, total urine arsenic level was positively associated with increased prevalence of type-2 diabetes. This finding supports the hypothesis that accumulative effect of even low level exposure to inorganic arsenic in drinking water may be responsible for prevalence of diabetes epidemic.

Despite the fact that global phase out of twelve most hazardous POPs, like PCB, Dioxins, and furan, began in 2004 under the Stockholm-Convension, a combination of factors have led their continued production, trade, use and release into the environment in India. OC and organophosphate (OP) pesticides are likely to pollute lipid containing food sources like milk owing to their lipophilicity. Recently presence of the high concentrations of BHC, aldrin, dieldrin, endrin and heptachlor has been reported from Malsinghwala village, Chandigarh (India). In a study on 147 samples of bovine milk collected from over 14 districts of Haryana (India), presence of HCH, DDT, endosulfan, and aldrin were detected in 100%, 97%, 43%, and 12% samples. Concentrations of ²-HCH and p,p'-DDE were more as compared to other isomers and metabolites of HCH and DDT. Presence of BHC (75%), methyl parathion (37.5%), dieldrin and 2,4,DDE (12.5% each) were also found in bovine milk samples collected from small scale rural and urban dairies of Allahabad region in Uttar Pradesh (India). Rural dairy samples carried higher loads of methyl parathion which may be attributed to the large scale agricultural practices in the area.

Indians vow by mothers' milk. Recently, Madras Diabetes Research Foundation, Chennai has advocated increase the practice of breast-feeding in children that helps prevent development of diabetes in future generation as well as in the feeding mothers by virtue of several hormonal, biochemical and physiological mechanisms. It is agonizing to state that pesticides have not spared even mother's milk. A study of mothers' milk from metropolitan city Chennai (Madras), Perugudi (situated at suburb of Chennai), Chidambaram (an agricultural town 250 Km South of Chennai) and Parangipettai (a fishing village 15 Km North of Chidambaram, Tamilnadu, India) revealed the presence of measurable concentrations of HCH, DDTs, PCBs, CHLs, and HCBs. Authors report that concentrations of OC pesticides HCHs and DDTs have increased in Chennai mothers milk in last decades. A recent study that analyzed contamination status of OCs in human milk across three major metropolitan cities, New Delhi, Mumbai and Kolkata found that presence of DDT was predominant followed by HCHs and PCBs. CHLs and HCB were noticed in lower concentrations. Based on estimated daily intake, this study observed that some infants were exposed to OCs to a greater extent, particularly HCHs than the guideline standards.

It is important to understand that metabolism of POPs in mammalian system is intractable and half-life of compounds range from 7 to 10 years in humans. Therefore, accumulative effect of even low dose exposure in general population over time and presence in the body may have long term detrimental effects. Dioxins and TCDDs are congener of POPs and have been reported to modify glucose metabolism, increase insulin resistance, cause hypoinsulinemia through an alteration of pancreatic membrane tyrosine phosphorylation, suggesting their involvement in the pathogenesis of type-1 and type-2 diabetes mellitus. Heavy presence of these POPs in bovine as well as human milk and its consumption in India may find link with altered glucose metabolism, increased incidence of insulin resistance in Indian adults as well as children population and relate to the increased incidence of diabetes for milk is an important essential diet in Indian homes be it urban or the rural. These finding may explain the quarry as to why the incidence of disease is soaring not only in the meat-addicted West but also in countries such as India, where many millions are vegetarian.

Arsenic contamination of ground water is a common phenomenon in the Bengal-Delta plain, covering the State of West Bengal, Jharkhand, Bihar, Uttar Pradesh, Assam and other Northern States of India. Recently from South India, ground water of several villages in Raichur district of Karnataka State has been found affected heavily with arsenic contamination. Table 1 presents arsenic contaminated districts and towns from different States of India. Arsenic consumption has been very common in the region of Chandigarh (Northern India) where prolonged consumption occur as a result of intake of arsenic contaminated water, vegetables, adulterated opium, Ayurvedic and indigenous medicines and home made brews. Increased use of fertilizers, insecticides and industrial activities also cause increased release of arsenic in Indian soil.

Table 1 : Cities and towns under which arsenic contaminated areas are explored in different states of India [Compiled from references 32-38]

States	Cities/Towns
Andra Pradesh	Medak, Pathancheru.
Arunanchal Pradesh	Dibang valley affected maximum including 6 other districts
Assam	Karimganj, Jorhat, Dhamji, Golaghat, and Lakhimpur affected maximum amongst the 21 affected districts.
Bihar	Bhagalpur, Munger, Buxer, Khagaria, Samastipur, Vaishali, Begusarai, Lakhisarai, Bhojpur, Patan, Saran (Chapra), Katihar, Purvi and Paschimi Champaran, Sitamarhi, Madhubani, Supaul, Araria, Kishanganj, Purnia.
Chhattishgarh	Rajanandgaon, Durg, Ambagarh chawki, Raipur.
Delhi	New Delhi
Jharkhand	Sahibganj
Karnataka	Raichur, Sindhanur, Lingsugur, Deodurg, Manvi, Maski.
Madhya Pradesh	Kaner, Betul.
Manipur	Thuobal.
Nagaland	Mon, Mokokchung.
Punjab	Patiala, Chandigarh.
Tripura	Dhalai, West and North Tripura Districts.
Uttar Pradesh	Lakhimpur-Kheri, Ballia, Kanpur Nagar, Bahraich, Balarampur, Gorakhpur, Chandauli, Ghazipur, Varanasi, Unnao, Faizabad, S. R. Nagar, Lucknow, Pilibhit, Shravasti, Siddharth Nagar, Maharaj ganj, Kushinagar.
West Bengal	Maldah, Murshidabad, North and South 24 Pargana, Nadia, Howrah, Hoogly, Bardhaman, Kolkatta, Kooch bihar, North and South Dinajpur.

Concentrations of and exposure to POPs and inorganic arsenic in many societies around the world may be different. Therefore, the risk of diabetes development could be graver than contemplated. Diabetes is a life long debilitating disease. It is noted in the beginning of the essay that the projected number of diabetics in India by 2025 would be around seventy million. If the exposure to arsenic contamination plays an important and causal role in diabetes development, nearly fifty million people living in 3200 villages alone in nine of the total eighteen districts of West Bengal, are affected with arsenic contaminated drinking water. Rice and vegetable fields irrigated by such ground water further receive large amount of arsenic in this area. Recently 30,000 people were found chronically exposed to arsenic contaminated water from a small town Maski alone in Karnataka.

Despite the fact that globalization and industrialization has contributed significantly in improving living standards of Indian, major population still depend on agricultural practices. Increasing the production of food grains and other commodities is mandatory for India to feed the world's second largest population. It is apparent therefore, that this goal could not be easily achieved without help of chemical fertilizers, insecticides, pesticides and digging the ground for irrigation bore-wells. Together, all these activities play role in exaggerating POPs and arsenic contaminations. There is lack of diabetes prevalence and incidence data from POPs and arsenic affected regions of India. In view of the statement of Shaukat Sadikot, President, Diabetes India, that four out of five people with diabetes in rural and tribal India do not know that they have it; it is quite possible that the real situation may be more grave than predicted. In the light of above developments, it becomes imperative to look beyond the conventional belief that diabetes in merely consequence of gluttony and laziness in this era. There is an urgent need to accelerate research in this area linking toxicological impacts of various pollutants, have realistic data at National level and educate people regarding the cumulative detrimental and long-term harmful effect of even low exposure of these pollutants later in the life.

REFERENCES

Acharya, S.K. 2006. Arsenic contaminated ground water from parts of Indian subcontinent: insights into influence of geological setting and bio-geochemical processes. In *Souvenir of the 31st annual convention and third International seminar on exploration Geophysics.* 6-12 November, Hyderabad, pp. 21-39.

Ana, N-C., Silbergeld, E.K., Roberto, P-B., Guallar, E. 2008. Arsenic exposure and prevalence of type-2 diabetes in US adults. *J. Am. Med. Assoc.*, **300**, 814-822.

Ana, N-C., Silbergeld, E.K., Streeter, R.A., Clark, J.M., Burke, T.A., Guallar, E. 2006. Arsenic Exposure and type-2 diabetes: a systematic review of the experimental and epidemiologic residence. *Environ. Heath Perspect.*, **114**, 641-648.

Balasubramanyam, M. 2008. One more reason for breast feeding-prevention of diabetes! *Curr. Sci.*, **95**, 1116-1118.

Brown, P. 2008. Trouble in store: Bed diet? Too little exercise? The root cause of diabetes. *New Scientist*, **200(2673)**, 36-39.

Cox, S., Niskar, A.S., Venkat Narayan, K.M., Marcus, M. 2007. Prevalence of self reported diabetes and exposure to organochlorine pesticides among Mexican Americans: Hispanic Health and Nutrition examination survey 1982-1984. *Environ. Health Perspect.*, **115**, 1747-1752.

DeVito, M.J., Birnbaum, L.S., Farland, W.H., Gasiewicz, T.A. 1995. Comparisons of estimated human body burdens of dioxin-like chemicals and TCDD body burdens in experimentally exposed animals. *Environ. Health Perspect.*, **103**, 820-831.

Devonathana, G., Subramanian, A., Someyaa, M., Sudaryantoa, A., Isobea, T., Takahashia, S., Chakraborty, P., Tanabea, S. 2009. Persistent organochlorines in human breast milk from major metropolitan cities in India. *Environmental pollution*, **157**, 148-154.

Ebner, K., Matsumura, F., Enan, E., Olsen, H. 1993. 2, 3, 7, 8-Tetrachlorobenzo-*p*-dioxin (TCDD) alters pancreatic membrane tyrosine phosphorylation following acute treatment. *J. Biochem. Toxicol.*, **8**, 71-81.

Gopal, N. A. 2007. Mansa Village is dying, while Government sleeps. Chandigarh Newsline. *cities.expressindia.com.*, down loaded Jan. 15, 2009.

Gupta, R. 2004. More arsenic. *Down To Earth*, **13(8)**, 26-36.

Heneriksen, G.L., Ketchum, N.S., Michalek, J.E., Swaby, J.A. 1997. Serum dioxin and diabetes mellitus in veterans of Operation Ranch Hand. *Epidemiology*, **8**, 252-258.

Henry, C.J., Lightowler, H.J., Newens, K., Sudha, V., Radhika, G., Sathya, R.M., Mohan, V. 2008. Glycemic index of common foods tested in the UK and India. *Br. J. Nutr.*, **99**, 840-845.

Hoskote, S.S., Joshi, S.R. 2008. Are Indians destined to be diabetic? *J. Assoc. Physicians India*, **56**, 225-226.

Hossain, P., Kawar, B., Nahas, M.E.I. 2007. Obesity and diabetes in developing world-A growing challenge. *N. Engl. J. Med.*, **356**, 213-215.

Joshi, S.R., Parikh, R.M. 2006. Family history and pedigree charting-a simple genetic tool for Indian diabetes. *J. Assoc. Physician India*, **54**, 437-439.

Kozhissery, D. 2008. Now arsenic in Karnataka. *Down To Earth*, **17(12)**, 14.

Lee, D-H., Lee, I-K., Jin, S-H., Steffes, M., Jacob, D.R. Jr. 2007. Association between serum concentrations of persistent organic pollutants and insulin resistance among non diabetic adults; Results from the National Health and Nutrition Examination survey 1999-2002. *Diabetes Care*, **30**, 622-628.

Lee, D-H., Lee, I-K., Song, K., Steffes, M., Tascano, W., Baker, B.A., Jacob, D.R. Jr. 2006. A strong dose-response relation between serum concentrations of persistent of organic pollutants and diabetes; results from the National Health and Examination Survey 1999-2002. *Diabetic Care*, **29**, 1638-1644.

Mohan, V., Sharp, P.S., Cloke, H.R., Burrin, J.M., Schumer, B., Kohner, E.M. 1986. Serum immunoreactive insulin responses to a glucose load in Asian Indian and European type-2 (non-insulin-dependent) diabetes patients and control subjects. *Diabetologia*, **29**, 235-237.

Mohler, M.L., He, Y., Wu, Z., Hwang, D.J., Miller, D.D. 2009. Recent and emerging antidiabetic targets. *Med. Res. Rev.*, **29**,125-175.

Narang, A.P.S. 1987. Arsenicosis in India. *Clinical Toxicology*, **25**, 287-295.

Rai, S.N., Thiagrajan, S., Ratnakumari, Y. 2008. Water pollution and health hazards. In *"Proceedings of National Seminar on Food Nutrition: Societal Importance"*. 21-22 November, (eds. S.N. Rai and A.P. Singh.) NGRI, Hyderabad. pp. 36-45.

Raji, A., Gerhard-Herman, M.D., Warren, M., Silverman, S.G., Raptopoulos, V., Mantzoros, C.S., Simonson, D.C. 2004. Insulin resistance and vascular dysfunction in nondiabetic Asian Indians. *J. Clin. Endocrinol. Metab.*, **89**, 965-3772.

Remillard, R.B., Bunce, N.J. 2002. Linking dioxins to diabetes: Epidemiology and biologic plausibility. *Environ. Health Perspect.*, **110**,853-858.

Roychowdhury, T., Tokunaga, H., and Ando, M. 2003. Survey of arsenic and other heavy metals in food composites and drinking water and estimate of dietary in take by the villagers from arsenic-affected area of West Bengal, India. *Sci. Total Enviorn.*, **308**, 15-35.

Sarkar, P. 2009. The dirty zones *www. Indiatogether org.* 2003. Dec. env-pops.htm downloaded Jan 15.

Schreinemachers, D.M. 2006. Mortality from Ischemic Heart disease and diabetes mellitus (Type-2) in four US wheat producing states: A hypothesis generating study. *Environ. Health Perspect.*, **114**,186-193.

Sharma, H.R., Kaushik, A., Kaushik, C.P. 2007. Pesticide residues in bovine milk from a predominantly agricultural state of Haryana India, *Environmental Monitoring and Assessment*, **129**,349-357.

Sicree, R., Shaw, J., Zimmet, P., Diabetes and impaired glucose tolerance. 2006. In: *Diabetes Atlas. International Diabetes Federation* (ed. Gan D, editor).3rd ed. Belgium. International diabetes Federation, pp. 15-103.

Singh, A.K. 2006. Chemistry of arsenic in ground water of Ganges-Brahmaputra river basin. *Cur. Sci.*, **91**, 1599-605.

Srivastava, S., Narvi, S.S., Prasad, S.C. 2008. Organochlorine and organophosphates in bovine milk samples in Allahabad region. *International J. Environ .Res.*, **2**, 165-168.

Subramanian, A., Ohtake, M., Kunisue, T., Tanabe, S. 2007. High levels of organochlorine in mothers' milk from Chennai (Madras) city-India. *Chemosphere,* **68**, 928-939.

Tiwari, A.K. 2005. Wisdom of Ayurveda in perceiving diabetes, enigma of therapeutic recognition. *Curr. Sci.*, **88**, 1043-1051.

Tiwari, A.K. 2008. Diabetic Epidemic: Consequences of wealth and luxury. *In 'Proceeding of National Seminar on Acharya Jagadish Chandra Bose and Ancient Indian Scientific Thought'* 21st December (ed. K V Krishna Murthy) Institute of Scientific Research on Vedas, Hyderabad. pp. 68-72.

Tseng, C-H., and Ana, N-C., Guallar, E. 2008. Arsenic exposure and diabetes mellitus in the United States *J. Am. Med. Assoc.*, **300**,2728-2729.

Walgate, R., Diabetes Research for Developing Countries. 2008. In *European Action on Global life Science (EAGLES), European Federation of Biotechnology. Diabetes Report.* pp.1-16.

Zimmet, P.Z., MacCarty, D.J., de Courten, M.P. 1997. The global epidemiology of non-insulin dependent diabetes mellitus and the metabolic syndrome. *J. Diabetes Complications*, **11**, 60-68.

❑❑❑

Chapter 21

Organic Livestock Production

L.S. JAIN

Professor (Retd) Animal Husbandry, Rajasthan College of Agriculture, MPUAT, Udaipur

Organic live stock production is one where right from birth the live stock are raised without modern veterinary medicines, growth promoters etc. the fodder fed should also be produced under organic farming rules are to be farmed to help those involved in the organic industry understand what is generally expected of them in producing , processing and marketing organic live stock products.

In this chapter basic introduction to organic livestock production i.e; out ling the steps that producers must take to obtain and retain organic farm certification and how pastures and animals must be managed. Steps are;

- Writing an organic plan: A farm management plan is the first step towards certified as a producer of organic live stock.
- Managing pasture land: Sustainable and productive pastures for livestock require close attention to forages and their effective use.
- Acquiring and caring for livestock: Humane treatment , nutritious feed and preventive health care are keys to successful production of organic livestock.
- The final rule: It provides complete text of the regulations that applies to organic livestock production.
- Recommended resources: These sources of information can expand a farmer`s education and understanding on the complex topic.

WRITING AN ORGANIC PLAN

Organic livestock production begins with writing an organic system farm management plan. For this the farmers must do the following:

- List the farmers production practices and procedures.

- Describe how these practices and procedures will be monitored .
- Explain the record –keeping system that documents organic production.
- Should prevent co- mingling of organic and non organic products.
- Record the composition and source of each substance used on the farm as a production input.
- Plan for future improvement.

As part of farm management plan, the organic farmers must draw an accurate in-scale map of the farm showing each production area and the acreage in each area depicting name and number of paddocks , pens and production building, field, boundaries, roads -ways, what the adjoining land is used for, how to protect organic production from outside influences and all sources of ground and surface water.

For organic dairy production farmers should fully describe the history of animal and crop area by listing what types of livestock and crops were grown in each area, what types of inputs such as feeds and fertilizers were used and whether all of these areas were involves in organic or conventional production.

MANAGING PASTURE LAND

The goal of pasture management includes ensuring tender, nutritious forage and effective forage use.

ORGANIC MATTER

Farmers can improve the general health of soil by boosting its organic matter content. Organic matter builds long term nutrient pool in the soil and create an efficient soil structure where nutrients and water flows freely to plants. It encourages populations of living organisms that keeps the whole biological process going.

Organic farmers must employ composting ,use of animal or green manures, forage legumes, crop rotation, rotational grazing and inter cropping for producing healthy crops and forage for livestock without purchasing commercial organic fertilizers and other amendments. For adjusting pH lime can be used. Similarly rock powders that contains nutrients not supplied by organic matter and humas.

Organic farmers must not use prohibited substances including, fertilizers to improve fertility such as ammonium nitrate, urea and super phosphates bio solids such as municipal sewage sludge and industrial by products containing heavy metals and or prohibited chemicals. They allowed fertilizers and amendments include plant materials like crop residues, compost, wood ash, rock dusts, granite dust lime and dolomite, animal by products such as blood meal, bone meal and feather meal, livestock manure including poultry litter and marine products like sea weed extract and fish meal and emulsion.

FORAGE SELECTION

Farmer should select forage crops that adopts to local soil and climatic conditions ,need less expenses and that re-grow easily, provide abundant growth palatable and succulancy and abundant nutrients, vitamins and minerals in sufficient amount and should provide good carrying capacity over the grazing period. The seeds used should not be genetically engineered or modified and should have not been treated with prohibited synthetic substances. Seeds should have been produced under organic standards.

Forage scarcity feeding the harvest as hay and silage. Pastures must be monitored frequently to evaluate its condition. Weeds, pests and diseases should not be controlled using prohibited substances . Crop rotation , rotational grazing ,nurse crops and natural practices can control these.

ACQUIRING AND CARING LIVESTOCK

Dairy animals purchased from conventional for marketing or using their milk or milk products as organic. For non organic for Poultry chicks obtained from non certified sources must be raised as organic from the second day of their life.

The producer of an organic livestock must not use; I) ANIMAL DRUGS INCLUDING HORMONESTO PRODUCE GROWTH, II) PROVIDE FEED SUPPLEMENTS or additives in amounts above those needed for adequate nutrition and health maintenance for the species at its specific stage of life, iii)feed plastic pellets for roughage ,iv)feed formula containing urea or manure and v)feed mammalian or poultry slaughter byproducts to mammals or poultry.

In so far as livestock health care is concerned the producer must (a) establish and maintain preventive type of livestock with regard to suitability for site specific conditions and resistance to prevent diseases and parasites, (b) provision of a feed ration sufficient to meet nutritional requirements including vitamins, minerals, protein and or amino acids ,fatty acids energy sources and fiber,(c) establishment of appropriate housing ,pasture conditions and sanitation practices to minimize the occurrences and spread of diseases and parasites. (d) provision of conditions which allows for exercise , freedom of movement and reduction of stress appropriate to the species. (e) administration of vaccines and other veterinary biologics and (f) make physical alterations as needed to promote the animals welfare and in a manner that minimizes pain and stress.

When preventive practices and veterinary biologics are inadequate to prevent sickness, a producer may administer synthetic medicines if such medicines are allowed under rules. The products of such animals that are on synthetic drugs and medicines shallnot be sold as organically produced.

❑❑❑

Chapter **22**

Farming Methods and Animal Health

BINAY KUMAR SINGH AND LIJO THOMAS

Scientist, Directorate of Rapeseed-Mustard Research, Bharatpur - 321 303, Rajasthan

INTRODUCTION

Production of organic food is founded on some basic principles defined by International Federation of Organic Agriculture Movements (IFOAM) and regulated by a set of definite rules. These principles do mention quality as desired aim but none of the specified requirements for qualifying for the label of organic food actually refer to it. In a strict sense, the quality of organic food is thus a process quality rather than a product quality. It is mainly, or only, indirect, results of the specified farming methods. Though, a number of systematically-controlled studies conducted in the past indicate that farming methods do make a difference to food composition or other aspects of its quality, many of these methods are not necessarily restricted to either organic or conventional systems. Thus, understanding the link between farming methods and food quality is imperative, to allow improvement of the products of any system, whether conventional or organic.

FARMING METHODS AND FOOD QUALITY

Tentatively, crop production can be described as a function of nutrient inputs (primarily nitrogen) and foreign compounds (pesticides). The two farming methods (conventional and organic) cannot be fully separated, though, the average farming conditions may vary. It is significantly important when we compare conventional versus organic food because if the conditions are kept identical, significant difference between the two product types cannot be expected. The influence of cultivar and soil type should also be taken into consideration when comparing crop products. In particular the production systems that tend to use different cultivars or procure raw materials produced in different locations have to be taken into account in a comparison, if the purpose is to determine the effect of consumer health, and/or satisfaction with sensory qualities of the food.

ASPECTS OF FOOD QUALITY

Quality of food is a collection of properties that satisfy or exceed the expectations of the consumer. It comprises a range of benefits: authentic, functional, biological, nutritional, sensual and ethical. However, any general method to measure the effect of food on the full range of benefits is lacking. Currently available literature covers only qualities relative to consumer health ie. functional, biological, nutritional and in some cases sensory properties when related to health. Some of the aspects of food-related health that are directly related to the chemical, microbiological and physical properties of the food are:

SAFETY FROM PATHOGENS

Available reports indicate that microbiological risk from organic food is smaller or similar to corresponding conventional foods. Few workers claim that organic food may be safer than estimated. Researchers attribute this quality of the organic food to the farming methods. For example, more extensive use of grass and other roughages in organic animal production improves the ability of the animals to eliminate zoonotic pathogens. Genetic resistance of the animals is yet another aspect that organic farmers bother when selecting breeding-stock. Although, studies regarding impact of animal genetic resistance on food safety is lacking, this is a potential area where organic production system developments can directly benefit conventional procedures.

As regards the transfer of pathogen to vegetables (from manure or other inputs of organic farming), no studies definitively rule out that pathogens can enter the organic food chain through animal or plant products. However, there are reports that support the observations made by organic farmers of relatively good resilience to infections by the organically grown plants. It is also reported that pathogens from organic animals are more susceptible to antibiotics and are easier to eliminate from patients.

SAFETY FROM TOXIC SUBSTANCES

Toxins may be intentionally applied or they may occur as pollutants in food. Synthetic pesticides, food additives and veterinary medicines are among the toxins which are added intentionally to the food either to increase its self life or productivity. On the other hand, dioxins, polychlorinated biphenyl (PCB) and dichlorodiphenyltrichloroethane (DDT) are some of the toxins unintentionally occurring in food as pollutants. These toxins, whether added intentionally or unintentionally to the food, are perceived as possible health risks and are thus unacceptable to the consumers. Detectable pesticide residues have been consistently found in a substantial proportion of samples analyzed from conventional farms. These residues are commonly not present on organic produce and, where present, residue levels are significantly lower corresponding to conventional ones. However, the occurrence of residues in present day conventional food is not considered a significant health risk because all pesticides, food additives etc. used by conventional food producers are thoroughly tested to determine whether they pose any health risk and if such indications are found, they are prohibited. But, at the same time it is also true that chemicals considered harmless today might be labelled 'poisonous' in the future, and little is known about how combinations of low levels of pesticides affect human health. Besides pesticides, food additives like preservative substances and food colourings used in conventional food are also considered as possible health risks, but here also, the effects are difficult to measure, and there is no consensus in the scientific community on safe levels.

So although there is no doubt that the presence of pesticide residues and food additives is significantly different between organic and conventional foods, there are no generally accepted data showing that the difference in composition makes a difference to the health of the consumers.

NATURALLY OCCURRING TOXINS-MYCOTOXINS

Mycotoxins are fungal metabolites produced in infected plant materials following infection with various fungal species, such as aflatoxins from *Aspergillus* species and fumonisins produced by strains of *Fusarium*. Studies of samples taken from fields have shown either no difference in mycotoxin content or lower content in the organic samples compared to the conventional ones. An experiment conducted in wheat under controlled cultivation revealed that the infection rate was consistently highest when only conventional mineral fertilizer was used, intermediate when both mineral fertilizer and pesticides were used and lowest in organic wheat grown with organic fertilizer and no pesticides. However, of note to the organic community is that although the relatively low and/or slow nutrient supply in organic farming helps to trigger the plant's natural defences against infection, it is achieved at the expense of some of the theoretical yield potential. Further, this problem cannot be solved by improving the methods for acquisition, retention and release of nutrients-excessive use of organic fertilizers may lead to the same quality problems that haunt conventional cereal producers.

In another study related to the concentration of mycotoxins in organic cereals, it was observed that improper management, drying and storage facilities are the major reasons. Nevertheless, this caused the average contamination in marketed foods to be higher in organic than in conventional.

Apart from the question of how the farming system effects concentrations of mycotoxin. Additional uncertainty exists in relation to the effects of different levels on human and animal health. Safety limits set too low may lead to the destruction of large volumes of food, whereas limits set too high expose consumers to significant health risks. As regards to mycotoxins in organic food, the recent increased acceptance of the concept of hormensis (ie. effects at low concentration can be opposite, normally beneficial, than the normally toxic effects at high doses) is likely to lead to a complete re-evaluation of the existing framework for safety evaluation, since 22 mycotoxins studies are reported to demonstrate hormensis. In such a case consumption of food with low levels of mycotoxins may strengthen the immune system rather than harming it, as evident by an *in-vitro* lymphocyte culture.

NATURALLY OCCURRING TOXINS - PLANT TOXINS

These are produced by plants as a defence against diseases and pests, including essential oils in myrtaceous plants and glucosinolates in *Brassica* species. Many of the plant toxicants are phytoalexins, which accumulate in varying amounts in response to mechanical damage, an infection or other stress conditions, so the concentration within the same plant often vary by more than a factor of 100 depending on plant age and sample site. Fortunately, humans and most herbivores have evolved the ability to degrade and/ or excrete most naturally occurring plant chemicals. As for now, there are no reports of long-term harmful effects on any humans of low to medium exposures to toxins from normal food plants, unless combined with malnutrition.

Nevertheless, in food plants such as potatoes, toxicants regularly reach such high levels that it can cause unpleasant symptoms to sensitive persons, although serious cases are extremely rare, and no cases have been reported in developed countries for more than two decades.

In terms of organic food quality, the question is then whether high, acutely toxic concentration occur in the relevant plant foods and if so, what production factors are important to predict and preferably prevent this from happening, or if other measures can be used to alleviate the harmful effects.

The existing data indicate that organically grown, undamaged material has somewhat higher average levels of these toxicants than conventionally grown crops. However, these values were well below the established safe limits. Recent study with parsnips showed equal levels in organic and conventionally grown, fresh undamaged material, but the increase after damage or storage was significantly lower for organic than conventional crops. However, studies showing contrasting results are also available.

Overall, no recent data indicate any particular risk of toxic symptoms to consumers from natural plant toxicants, and some hypotheses, if they are confirmed, could point to organic growing conditions as a measure that could further reduce the risk. However, since the most effective protection is simply to stop eating a dish if it makes one feel sick, the issue of organic production methods is of little direct significance for this aspect of food quality.

MEASUREMENT OF THE EFFECTS OF FARMING METHODS ON FOOD QUALITY

Almost all the reported studies pertaining to effect of farming methods on health have been performed on animals and until now, the results of these studies have not been particularly definitive. Though, some of the animal feeding studies do show significant effects on one or the other health related aspects, these effects are not always consistent. In the recent past, some workers have identified certain definitive markers of the effect of production methods on health. These are as following:

1. Biomarkers for protein oxidation - These markers are sufficiently sensitive to detect differences in human caused by small differences in diet.
2. Lymphocyte Proliferative Capacity - This study can differentiate between the effects of completely matched diets only differing in production factors. However, this marker is generally less applicable since it requires the sacrifice of the animal.
3. Sleep patterns of animals
4. Content of immunoglobulin A (IgA) in blood
5. Accumulation of adipose tissue

All the above markers are equally relevant to humans as these can be assessed non-invasively, except lymphocyte proliferative capacity.

The effect of farming methods on health can also be measured indirectly by measuring the content of beneficial substances in the food. Most studies in this regard have been done by analysing the nutrient content and assuming that it would somehow be possible to translate these data into a ranking of effect on health. Some of these studies have been reviewed in detail and following conclusions have been drawn:

The content of compounds indicated as positive for health in organically grown plants tend to differ from those provided with large amounts of easily available mineral fertilizers. Some of the measurable components are:

1. Higher level of ascorbic acid(vitamin C)
2. Lower levels of nitrate
3. Lower levels of total N(Protein)
4. Higher proportion of essential aminoacids in protein
5. Higher Zinc to phytate ratio
6. Lower levels of β–carotene and
7. Higher levels of plant secondary metabolites

Correspondingly, products from animal fed a large proportion of grass or other relevant roughage have higher:

1. Levels of β- carotene and other carotenoids, such as lutein
2. Proportion of conjugated linoleic acids in the fat
3. Proportion of polyunsaturated fatty-acids in the fat, and
4. Levels of vitamin E

It may be unexpected that the nitrates and the plant secondary metabolites are indicated as beneficial compounds, since nitrates are traditionally considered harmful irrespective of source and many secondary metabolites are known as toxicants. However, nitrate in vegetables is not associated with any diseases and may have direct, positive effects on human and animal health by protecting against harmful bacteria. Similarly, there are evidences to show that secondary plant metabolites exert a variety of physiological effects, which makes it likely that at least some of these effects (particularly the antioxidant property) are positive, and outweigh other, negative effects.

CONCLUSIONS

There have been very few scientific studies in which foods grown conventionally have been compared, under comparable and controlled conditions, with those produced organically, in terms of their nutrient composition or their biological effects on animals or human subjects. It would appear that few differences can be demonstrated, and where differences are detected they are very small. It is unlikely that the small differences in the nutrient content would have health implications for consumers. Future studies that can evaluate effects of growing conditions on levels of protective photochemical are urgently required.

Animal studies that have been conducted to compare health effects of conventionally- and organically- grown produce are very limited in number and have involved only a few species. Limitations in the study design and lack of internal consistency in the findings mean that the present data are of limited value in providing a scientific basis for evaluating health effects of organic foods. There are virtually no studies of any size that have evaluated effects of organic versus conventionally- grown foods on human health. Observational studies that have been conducted to compare the health profiles of organic and conventional farmers are of questionable value because of the possibility of confounding by other lifestyle factors.

The quality and quantity of the science applied in this area is inadequate. Conclusions cannot be drawn regarding potentially beneficial or adverse nutritional consequences, to the consumer, or increased consumption of organic foods.

REFERENCES

Baker, B.P., Benbrook, C.M., Groth, E. and Benbrook, K.L. 2002. Pesticide residues in conventional, integrated pest management (IPM)-grown and organic foods: insights from three US data sets. *Food Addit Contam*. 19(5):427-46.

Bateman, B., Warner, J.O., Hutchinson, E., Dean, T., Rowlandson, P., Gant, C., Grundy, J., Fitzgerald, C. and Stevenson, J. 2004. The effects of a double blind, placebo controlled, artificial food colourings and benzoate preservative challenge on hyperactivity in a general population sample of preschool children. *Arch Dis Child*. 2004 89(6):506-11.

Birzele, B., Meier, A., Hindorf, H., Krämer J. and Dehne H.W. 2002. Epidemiology of *Fusarium* infection and deoxynivalenol content in winter wheat in the Rhineland, Germany. *European Journal of Plant Pathology*. 108:667-673.

Bourn, D. and Prescott, J. 2002. A comparison of the nutritional value, sensory qualities, and food safety of organically and conventionally produced foods. *Critical Reviews in Food Science and Nutrition*. 42(1):1–34

Couzin, J. 1998. Cattle diet linked to bacterial growth. *Science*. 281:1578-1579.

Diehl, J. F. 2002. Some established facts and some new concepts of food toxicology- a review. *Acta alimentaria*. 31:355-369.

Diez-Gonzalez, F., Callaway, T.R., Kizoulis, M.G. and Russell, J.B. 1998. Grain feeding and the dissemination of acid-resistant Escherichia coli from cattle. *Science*. 281:1666-8.

Elmholt, S. 2003. Ecology of the ochratoxin A-producing *Penicillium verrucosum* : occurrence in the field soil and grain with special attention to farming system and on-farm drying practices. *Biological Agriculture and Horticulture*. 20:311-337.

John, P., Morrissey and Anne, E. 1999. Osbourn fungal resistance to plant antibiotics as a mechanism of pathogenesis. *Microbiology and Molecular Biology Reviews*, 63(2):708-724.

Kouba, M. 2003. Quality of organic animal products. *Livestock Production Science*. 80:33-40.

Magkos, F., Arvaniti, F. and Zampelas, A. 2003. Organic food: nutritious food or food for thought? A review of the evidence. *Int J Food Sci Nutr*. 54(5):357-71.

❑❑❑

Index

❑❑❑